THE
DINOSAUR
HUNTERS

Also by Deborah Cadbury

The Feminisation of Nature

THE DINOSAUR HUNTERS

A Story of Scientific Rivalry
And the Discovery of the
Prehistoric World

Deborah Cadbury

FOURTH ESTATE • London

First published in Great Britain in 2000 by
Fourth Estate Limited
6 Salem Road
London W2 4BU
www.4thestate.co.uk

ISBN 1-85702-959-3

Typeset by Phoenix Typesetting, Ilkley, West Yorkshire
Printed in Great Britain by The Bath Press Ltd, Bath

For my mother and Martin,
the first readers,
with love

Contents

Picture Credits

p.11 The Natural History Museum, London; p.18 Portrait of William Buckland (1784–1856) Professor of Mineralogy at Oxford University and Dean of Westminster (oil on canvas) by Samuel Howell, Corpus Christi College, Oxford UK/Bridgeman Art Library; p.28 Reproduced with kind permission of the Geological Society; p.31 Mary Anning (1799–1847) (oil on canvas) by English School (19th century) Private collection/Bridgeman Art Library; p.34 By kind permission of The Natural History Museum, London; p.37 By permission of the President and Council of the Royal Society; p.44 By kind permission of The Natural History Museum, London; p.59 By kind permission of the Geological Society; p.60 Hulton Getty Picture Collection; p.66 By kind permission of the Geological Society; p.71 The Natural History Museum, London; p.73 The Deluge, 1834 (oil on canvas) by John Martin (1789–1854) Yale Center for British Art, Paul Mellon Collection, USA/Bridgeman Art Library; p.75 By kind permission of the Geological Society; p.83 By kind permission of the Geological Society; p.91 By kind permission of the Geological Society; p.107 By kind permission of the Geological Society; p.120 By permission of the President and Council of the Royal Society; p.134 Hulton Getty Picture Collection; p.137 Hulton Getty Picture Collection; p.150 By kind permission of the Geological Society; p.159 Alexander Turnbull Library, Wellington, New Zealand; p.162 Duria antiquior (ancient Dorset) depicting an imaginative reconstruction of the life of the Jurassic seas, engraved by George Scharf (1820–95) printed by Charles Joseph Hullmandel (1989–1850) (engraving) by H.T. De la Beche (19th century) Oxford University of Natural History, UK/Bridgeman Art Library; p.177 Courtesy of the Sussex Archaeological Society; p.189 Mary Evans Picture Library; p.191 By kind permission of the Geological Society; p.208 The Natural History Museum, London; p.209 Alexander Turnbull Library, Wellington, New Zealand; p.236 By kind permission of the Geological Society; p.248 By permission of the President and Council of the Royal Society; p.259 Hulton Getty Picture Collection; p.277 Hulton Getty Picture Collection; p.281 Hulton Getty Picture Collection; p.294 The Illustrated London News Picture Library; p.297 The Illustrated London News Picture Library; p.306 Hulton Getty Picture Collection; p.313 Hulton Getty Picture Collection; p.320 Hulton Getty Picture Collection.

Extracts taken from *The Journal of Gideon Mantell: Surgeon and Geologist* edited with an introduction and notes by E. Cecil Curwen (1940), reproduced with kind permission of Oxford University Press.

Acknowledgements

In writing this book, I am indebted to many specialists for their generous assistance with my research. I would particularly like to thank historian of science Professor Hugh Torrens at the Department of Earth Sciences, University of Keele. In addition to many fascinating discussions on Gideon Mantell, Richard Owen and Mary Anning, I am grateful to Hugh Torrens for giving up valuable time to read the manuscript and for offering expert advice.

At the Natural History Museum in London I would like to thank Dr Angela Milner and Sandra Chapman at the Department of Palaeontology for information on the evidence available to the early geologists and for many helpful insights into the history of palaeonotology. Thank you, too, to John Cooper at the Booth Museum of Natural History in Brighton for allowing me to use his archive collection revealing Mantell's fate in the 1830s; to Dr Joan Watson at the University of Manchester for information on fossil botany; to Dr David Norman at the University of Cambridge, and to numerous others who have helped with my research. Any remaining errors are my responsibility.

I owe a great deal to the studies of many other scholars cited in the references, and especially to the late John Thackray at the Natural History Museum in London who advised me on Richard Owen's key articles. The project was also ably supported by the archivists and librarians at the Royal Society of London, the Geological Society, Oxford University Museum, the Crystal Palace Museum and the Muséum National d'Histoire Naturelle in Paris. Thank you to the publishers, John Murray, for allowing me to cite from the works of Richard Owen and William Buckland, and from their biographies; to the Sussex Archaeological Society for kind permission to quote from Gideon Mantell's unpublished diary; and to John Wennerbom for permitting me to consult his excellent unpublished thesis: 'Charles Lyell and Gideon Mantell, 1821–1852: their quest for elite status in English geology'.

At Fourth Estate, I owe a debt of gratitude to Christopher Potter for sharing my vision of the possibilities for the narrative and for his skilled editorial judgement at every stage. Leo Hollis at Fourth Estate has been a wonderful support, providing expert guidance on each chapter and seeing the manuscript through to its final version. At Curtis Brown, Jane Bradish Ellames provided valued advice and encouragement for the project over many months.

Finally, special thanks are due to Julia Lilley for being a pillar of strength throughout and for reading and commenting on all the chapters, and to Martin Surr for his excellent judgement on many issues to do with the writing, and for sharing my enthusiasm for the emerging story.

PART ONE

I

An Ocean Turned to Stone

She sells sea-shells on the sea-shore,
The shells she sells are sea-shells, I'm sure
For if she sells sea-shells on the sea-shore
Then I'm sure she sells sea-shore shells.

Tongue-twister by Terry Sullivan, 1908,
associated with Mary Anning

O n the south coast of England at Lyme Regis in Dorset, the cliffs
tower over the surrounding landscape. The town hugs the coast
under the lee of a hill that protects it from the south-westerly wind. To
the west, the harbour is sheltered by the Cobb, a long, curling sea wall
stretching out into the English Channel – the waves breaking ceaselessly
along its perimeter. To the east, the boundary of the local graveyard
clings to the disintegrating Church Cliffs, with lichen-covered grave-
stones jutting out to the sky at awkward angles. Beyond this runs the
dark, forbidding crag face of Black Ven, damp from sea spray. The landscape then levels off across extensive sweeps of country, to where the
cliffs dip to the town of Charmouth, before rising sharply again to form
the great heights of Golden Cap.

At the beginning of the nineteenth century, according to local folklore, the stones on Lyme Bay were considered so distinctive that
smugglers running ashore on 'blind' nights knew their whereabouts just
from a handful of pebbles. However, it was not only smugglers and
pirates who became familiar with the peculiarities of these famous cliffs.

Through a series of coincidences and discoveries Lyme Bay soon became known as one of the main areas for fossil hunting. Locked in the layers of shale and limestone known as the 'blue lias' were the secrets of a vast, ancient ocean now turned to stone, the first clue to an unknown world.

In 1792, war erupted in Europe and it became dangerous for the English gentry to travel on the Continent. Many of the well-to-do classes adopted the resorts of the south coast of England. The dramatic scenery around Lyme Bay became a favourite among those who spent part of the season at Bath. In the summer, smart carriages often lined the Parade and the steep, narrow streets that nestled into the hillside. The novelist Jane Austen was among those who visited early in the nineteenth century. She was charmed by the High Street, 'almost hurrying into the sea', and 'the very beautiful line of cliffs stretching out to the east'. The Cobb curving around the harbour became the dramatic setting for scenes in her new novel *Persuasion*. It was here that Louisa Musgrove fell 'lifeless . . . her eyes closed, her face like death', and was nursed back to health by the romantic sea captain.

Jane Austen's letters to her sister, Cassandra, reveal that during her short stay she met an artisan in the town by the name of Richard Anning. He was summoned to value the broken lid of a box and, according to Jane Austen, was a sharp dealer. She told her sister that Anning's estimate, at five shillings, was 'beyond the value of all the furniture in the room together'.

Richard Anning, even as a skilled carpenter, struggled to make a living. The blockade of European ports during the Napoleonic Wars had caused severe food shortages. With no European corn available, the price of wheat had risen sharply, from 43 shillings a quarter in 1792 just before the war, to 126 shillings in 1812. Since bread and cheese was the staple diet for many in the southern counties, the spiralling price of a loaf caused great suffering. Wages did not rise during this period, and in many districts workers received a supplement from the parish to enable them to buy bread. Industrious labourers effectively became paupers relying on parish charity, and there was a real fear of starvation. While the gentry, glimpsed beyond sweeping parklands in their country estates, benefited from high prices and seemed impervious to the effects of war, the poor began to riot. The flaming rick or barn became a symbol of the times. Richard Anning was himself a ringleader of one protest over food shortages.

In rural Dorset, the poor were not only hungry, but with a shortage of fuel they also faced damp, cold conditions and sometimes worse. Richard Anning and his wife, Molly, lived in a cottage in a curious array of houses built on a bridge over the mouth of the River Lym. On one occasion, they awoke to find that 'the ground floor of their home had been washed away during the night'. Their modest home had succumbed to an 'exceptionally rough sea which had worked the havoc'.

The desire to keep warm could have lain behind a tragedy that befell the Annings' eldest child, Mary, at Christmas in 1798. The event was reported starkly in the *Bath Chronicle*: 'A child, four years of age, of Mr R. Anning, a cabinet maker of Lyme, was left by the mother about five minutes . . . in a room where there were some shavings by a fire . . . The girl's clothes caught fire and she was so dreadfully burnt as to cause her death.' Whether Mary was huddling too close to the flames for warmth, or accidentally stumbled, is not known. It is known, however, that her distraught mother, on the birth of their next daughter six months later, called her Mary in memory of her dead sister.

Naming a newborn after a child that had died was a common practice at a time when a quarter of poor infants died in their first year and half were dead before the age of five. Many were undernourished and readily succumbed to consumption, pneumonia, smallpox, measles or other diseases. Apart from the sudden death of their eldest daughter Mary, the Annings had already lost two other children, Martha and Henry, by the year 1800. But fate was to intervene in an unexpected way in the young life of the second Mary Anning.

That summer, when Mary was just one year old, news reached Lyme Regis that a touring company of riders was to perform near the town. Among the enticements were a display of vaulting, riding stunts and a lottery, with prizes such as copper tea-kettles and legs of mutton. The arrival of the travelling performers was a welcome distraction for the local inhabitants, and crowds of people trekked past the church and the gaol near the Annings' house to the equestrian show, set in a field on the outskirts of town. Mary was taken along in the care of a local nurse, Mrs Elizabeth Hasking.

By late afternoon a heavy thunderstorm developed, but the crowds would not disperse, perhaps lingering to see who had won the lottery. Then, in the words of the local schoolmaster, George Roberts: 'a vivid discharge of electric fluid ensued, followed by the most awful clap of thunder that any present ever remembered hearing, which re-echoed around the fine cliffs of Lyme Bay. All appeared deafened by the crash. After a momentary pause a man gave the alarm by pointing to a group that lay motionless under a tree.'

There were three dead women, among them Mary's nurse, Elizabeth, whose hair, arm and cap along the right side were 'much burnt and the flesh wounded'. She was still holding the baby, who was insensible and could not be roused. The second Mary Anning, known to be 'dear to her parents', was carried back to Lyme, 'in appearance dead'. But when bathed in hot water, gradually she was revived, to the 'joyful exclamations of the assembled crowd'. According to the family, this was a turning-point for the young Mary Anning: 'She had been a dull child before but after this accident she became lively and intelligent.'

As Mary grew older, she took a keen interest in helping her father gather fossil 'curios' from the beach to sell to tourists. In the early part of the century, Richard Anning had several more children to support: the boys Joseph, Henry, Percival and Richard and another daughter, Elizabeth. To supplement his meagre income as a carpenter, Mary and her father set up a curiosity table outside their home to sell their wares to the tourists. However, selling fossils was a competitive business.

One collector, called the 'Curi-man' or Captain Cury and known locally as a 'confounded rogue', would intercept the coaches and sell specimens to travellers on the Exeter to London turnpike. Another ill-fated collector was Mr Cruikshanks, who could often be seen along the shoreline with a long pole like a garden hoe. When Cruikshanks lost the small stipend supporting him, leaving nothing but a tiny income from the sale of curios, he closed the account of his miserable existence and committed suicide by leaping off the Gun-Cliff wall in the centre of Lyme into the sea.

No one could explain what these 'curios' were. Petrified in the rocks on the shore were strange shapes, like fragments of the backbone of a giant, unknown creature. These were sold locally as 'verteberries'. There were enormous pointed teeth, thought to be derived from alligators or crocodiles. Relics of 'crocodilian snouts' had been reported in the region for several years. There were also pretty fossil shells and stones, called 'John Dory's bones' or 'ladies' fingers'.

At the time, throughout England, superstitions abounded about the meaning of fossils. The beautiful ammonites, called 'cornemonius' in the local dialect, with their elegant whorls like the coils of a curled-up serpent, were also known as 'snake-stones'. The subject of the wildest speculation, such stones were thought in earlier centuries to have magical powers, and could even serve as an oracle. The ammonite, it was believed, could bring 'protection against serpents and be a cure for blindness, impotence and barrenness'. Occasionally a snake's head would be painted on the coils to be used as a charm. But snake-stones were not always a symbol of good fortune. In some regions it was thought that they were originally people, who for their crimes were first turned into snakes and then cast into stone. By divine retribution anyone who was evil could be turned to dust, just as Lot's wife had been turned into a pillar of salt.

There were other strange curios, too, such as the long, pointed belemnites. These were said to be thunderbolts used by God, known colloquially as 'devil's fingers' or 'St Peter's fingers'. These also had special powers. According to ancient tradition powdered belemnites could cure infections in horses' eyes, and water in which belemnites had been dipped was even thought to cure horses of worms.

The fossils that resembled fragments of real creatures like snakes or crocodiles defied explanation. Myths of the time give tantalising

Lithograph (1825) of the Cobb at Lyme in which
the figure is thought to be Mary Anning.

insights. Some held that they were the 'seed' or 'spirit' of an animal, spontaneously generated deep within the earth, which would then grow in the stone. According to others, fossils were God's interior 'ornament' of the earth, just as flowers were the exterior ornament. They might even have been planted by God as a test of faith! After all, if they were the remains of real animals that had once thrived, how had they burrowed their way down so deep into the rocks? And why would any creature do this? Alternatively, if the rocks had formed gradually around them, long after the animals had perished, this implied that God's Creation had occurred over a period of time, not in a few days as described in Genesis. Entombed in the stony cliff-face was a mystery beyond explanation.

At the beginning of the nineteenth century many had absolute faith in the word of the Bible. To them, the most convincing explanation was that these were the remains of creatures that had died during Noah's Flood and had been buried as the earth's crust re-formed. Although there are no records of Mary Anning's view as a child, it seems likely that this was the framework of colourful folklore and unyielding religious belief that informed her searches along the cliffs of Lyme Bay.

Mary became skilled at searching for 'crocodiles'. Laid out on the table before their house were giant bones of 'Crocodiles', 'Angels' Wings', 'Cupid's Wings', 'Verteberries', and 'Cornemonius'. Her searches on the beach made her mother Molly Anning very angry, as, according to Roberts the schoolmaster, 'she considered the pursuit utterly ridiculous'. It was also dangerous. Rainwater endlessly percolating through layers of soft shales and clays caused frequent mudslides and rockfalls, especially in winter. There was also the risk of being caught by the sea as the fossils, revealed by erosion, had to be removed before the tide turned and the waves washed them away. Sometimes Mary and her father were trapped by the rising waves between the sea and the cliffs, and had to struggle up the slippery rockface to safety. On one occasion, Richard Anning was caught in a landslide as part of the Church Cliffs collapsed into the sea, and narrowly escaped being carried down with the rocks and crushed on the beach below.

One night in 1810, however, Anning was not so lucky when, taking a short cut to Charmouth, he strayed from the path and fell over the treacherous cliffs at Black Ven. He was severely weakened by his injuries and soon succumbed to the endemic consumption and died. Molly and the children were destitute. They had no savings; indeed, Richard Anning had left his family with £120 worth of debt, a large sum at a time when the average labourer's wage was around 10 shillings a week. There was no way that Molly could readily pay back such a debt. As a result, she was obliged to face the humiliating prospect of appealing for help from the Overseers of the Parish Poor. It was a considerable misfortune for an artisan family.

Under the old Poor Laws dating from Tudor times, the poverty-stricken could be accommodated in one of fifteen thousand Poor Houses in England, where inmates struggled with conditions recognisable from the pages of Charles Dickens. Alternatively the poor received 'outdoor poor relief', as in the case of the Annings, which enabled them to stay in their own home while receiving a supplement from the parish. Although conditions on outdoor relief varied across districts, it was usually a miserly amount for food and clothing, or sometimes given in kind as bread and potatoes. The average weekly payment on outdoor poor relief was three shillings at a time when the minimum needed to scrape a living was six or seven shillings a week. Paupers were thus dependent on charity or could appeal to relatives for support. Older children were expected to help out with any number of tasks – horse holding, running as messengers, and cleaning or other domestic work. It was common for those on poor relief to be severely malnourished, and the hardships the Anning family endured were so severe that of all the children, only Mary and Joseph were to survive.

While Joseph, Mary's elder brother, was apprenticed to an uphol-sterer, Mary continued to search the beach for fossils. One day she found a beautiful ammonite, or snake-stone. As she carried her trophy from the beach a lady in the street offered to buy it for half a crown. For Mary this was wealth indeed, enough to buy some bread, meat and possibly tea and sugar for a week. From that moment she 'fully determined to go down upon the beach again'.

During 1811 – the exact date is not known – Joseph made a remark-able discovery while he was walking along the beach. Buried in the shore below Black Ven, a strange shape caught his eye. As he unearthed the sand and shale, the giant head of a fossilised creature slowly appeared, four feet long, the jaws filled with sharp interlocking teeth, the eye sockets huge like saucers. On one side of the head the bony eye was en-tire, staring out at him from some unknown past. The other eye was damaged, deeply embedded in the broken bones of the skull. Joseph immediately hired the help of two men to assist him and uncovered what was thought to be the head of a very large crocodile.

Joseph showed Mary where he had found the enormous skull, but since that section of the beach was covered by a mudslide for many months afterwards it was difficult to look for more relics of the creature.

Nearly a year elapsed before Mary, who was still scarcely more than twelve or thirteen, came across a fragment of fossil buried nearly two feet deep on the shore, a short distance from where Joseph had found the head.

Working with her hammer around the rock, she found large vertebrae, up to three inches wide. As she uncovered more, it was possible to glimpse ribs buried in the limestone, several still connected to the vertebrae. She gathered some men to help her extract the fossils from the shore. Gradually, they revealed an entire backbone, made up of sixty vertebrae. On one side, the shape of the skeleton could be clearly seen; it was not unlike a huge fish with a long tail. On the other side, the ribs were 'forced down upon the vertebrae and squeezed into a mass' so that the shape was harder to discern. As the fantastic creature emerged from its ancient tomb they could see this had been a giant animal, up to seventeen feet long.

News spread fast through the town that Mary Anning had made a tremendous discovery: an entire connected skeleton. The local lord of the manor, Henry Hoste Henley, bought it from her for £23: enough to feed the family for well over six months.

The skull of the unknown beast found by Joseph Anning in 1811,
now in the Natural History Museum, London.

The strange creature was first publicly displayed in Bullock's Museum in Piccadilly in the heart of London. It quite baffled the scholars who came to visit, as there was no scientific context in England within which they could readily make sense of the giant fossil bones. Geology was in its infancy and palaeontology did not exist. The peculiar 'crocodile', with its jaw set in a disconcerting smile and its enormous bony eyes, was something inexplicable from the primeval world. In the words of a report in Charles Dickens's journal, *All the Year Round*, there was to be a 'ten year siege before the monster finally surrendered' and revealed its long-buried secrets to the gentlemen of science. Nearly a decade was to elapse before the experts could even agree on a name for the ancient creature.

As news of Mary Anning's discovery reached scholarly circles in London and beyond, one of the first to visit her at Lyme Regis was William Buckland, a Fellow of the prestigious Corpus Christi College at Oxford University. Engravings of William Buckland portray a serious man, with even features and a broad expanse of forehead. Invariably, in these period poses, he is holding some fossil and formally attired in sombre black academic robes, looking the epitome of the nineteenth-century scientist. To those who knew him, he was renowned for qualities other than this stern and imposing image.

'Dr Buckland's wonderful conversational powers were as incommunicable as the bouquet of a bottle of champagne,' wrote Storey Maskelyne, one of his Oxford colleagues. 'It was at the feast of reason and the flow of social and intellectual intercourse that Buckland shone. A merrier man within the limit of becoming mirth I never spent an hour's talk withal. Nothing came amiss with him from the creation of the world, to the latest news in town . . . In build, look and manner he was a thorough English gentleman, and was appreciated within every circle.'

Although Buckland had a wide range of interests his greatest passion was for 'undergroundology', as he called the new subject of geology. Many of his holidays from Oxford were spent at Lyme, where he explored the cliffs 'with that geological celebrity, Mary Anning, in whose company he was to be seen wading up to his knees in the sea, searching for fossils in the blue lias'. At his lodgings by the sea, Buckland's breakfast table was 'loaded with beefsteaks and Belemnites, tea and Terebratula, muffins and Madrepores, toast and Trilobites, every table and chair as well as the floor occupied with fossils and rocks, earth, clays and heaps of books, his breakfast hour being the only time that the collectors could be sure of finding him, to bring their contributions and receive their pay'.

Born in the village of Axminster six miles inland from the Dorset coast, Buckland was no stranger to the impressive cliffs at Lyme. Since his childhood, the rocks of this region had enchanted him. 'They were my geological school,' he wrote, 'they stared me in the face, they wooed me and caressed me, saying at every turn, Pray, Pray, be a geologist!' His father, the Reverend Charles Buckland, had encouraged his enquiring approach to natural history. Following an accident, Charles Buckland was blind for the last twenty years of his life, but together father and son had explored the local quarries, the young William describing every detail of the beautiful fossil shells that his father could only touch. The boy's exceptional 'talent and industry' were noted by his uncle, a Fellow at Oxford University, who steered William's education, first to Winchester and then on to Corpus Christi College.

When William Buckland descended from his carriage in the city of famous spires at the turn of the nineteenth century, he had soon found that the university was steeped in an Anglican tradition in which the Scriptures, for many, were the key to understanding our history, and fossils were interpreted in this context. Most of the college lecturers took Holy Orders and advancement was principally through the Anglican Church. Buckland was himself ordained in 1809 and elected a Fellow in the same year.

At the time, more than a hundred years before radiometric dating was to dispel any lingering doubts about the vast antiquity of the globe, it was impossible to prove with certainty its exact age. For over two centuries, leading scholars had tried to solve this puzzle by taking the

Bible as evidence. Studies of the earth were carried out by classicists, who could analyse sacred writings in Hebrew, Latin or Greek. In 1650 the Archbishop of Armagh, James Ussher, had concluded that God created the earth the night preceding Sunday 23 October, 4,004 years before the birth of Christ. His calculation had been made by adding together the life spans of the descendants of Adam, combined with knowledge of the Hebrew calendar and other biblical records. His dating of the earth, far from being ridiculed, was accepted as an excellent piece of historical scholarship, and following his lead, the study of chronology using sacred texts became an established approach for the next two hundred years.

Other methods of dating the earth were occasionally put forward. In 1715, Edmond Halley had proposed an ingenious experiment to the Royal Society in which the rate of increase in the saltiness of lakes and oceans could be calculated, assuming that they contained no salt when the globe was created. However, his ideas were not pursued, and Halley himself thought his results were likely to confirm 'the evidence of the Sacred Writ, [that] Mankind has dwelt about 6,000 Years'.

Apart from revealing the age of the earth, the Bible had other geological implications that were to prove equally challenging for the early geologists like William Buckland. The prophet Moses outlined the story of Creation in which God made the Heavens, the Earth and every living thing in just seven days. In the biblical Creation story all creatures were made simultaneously. There is no prehistory in the Bible, and no prehistoric animals.

Moses also described a universal Flood in which 'all the fountains of the great deep and the windows of heaven were opened', and the entire face of the earth was wiped out, destroying all creatures except the few saved in Noah's Ark. Sacred texts were scrutinised so as to shed more light on these events. One highly respected seventeenth-century naturalist, a German Jesuit, Athanasius Kircher, produced a detailed paper on the dimensions of the Ark and its animal contents. This approach was still flourishing in 1815, when the Reverend Stephen Weston studied changing place-names in Hebrew and Greek and claimed to locate the very site where Noah's Ark came to land – on one of the highest mountains of the earth in Tibet.

At Oxford, William Buckland knew that anomalies unearthed in the rocks during the eighteenth century had challenged religious scholarship. Many stones resembling creatures or plants had been uncovered in locations that defied explanation. How could it be that sea shells were found on the peaks of the highest mountains? Was this evidence for the Flood and, if so, how had such vast amounts of water been suddenly generated and then fallen away? Savants were hard-pressed to explain why stones that looked just like animal teeth were found deeply embedded in solid rock, or how plants had become petrified within layers of coal. If fossils were the remains of animals, why were bones of tropical animals found in cold northern regions? Had the climate been mysteriously inverted? Stranger still, why was it that fossils resembling fish buried in one rock could be covered by layers of rock that contained only land animals, and in turn have shells and sea plants in the rocks above? This seemed to provide evidence of astonishing disorder and devastation, which was hard to understand if the world was purposefully designed in seven days by the Almighty Creator.

By the late eighteenth century scholars were making progress in understanding the history of the earth, not by taking the Bible as evidence, but the *rocks* themselves. One of the spurs for this was the growth of the mining industry in parts of Northern Europe such as Thuringia and Saxony. It was here on the present border between Germany and Poland that a pioneering thinker, Abraham Werner, created an order out of the seemingly haphazard formation of rocks beneath the earth's surface.

Abraham Werner was taken out of school at Bunzlau when his mother died, and sent to work for his father who managed the local ironworks for the Duke of Solm. He later entered the great Mining Academy of Freiberg, where his teaching on mineralogy became famous throughout Europe. Werner's ideas and others' showed that the earth's crust could be classified into four distinct categories of rock, which were always found to be in the same order of succession. The oldest of these were the crystalline rocks such as granite, gneiss and schist, containing no fossils. These became known as the *Primary* rocks, corresponding to the most primitive period of the earth's history, since these rocks were laid down first in the earth's crust. Above these in order of succession were the *Transition* rocks, including greywackes, slates and limestone. Only a small number of fossils could be found here. This was followed by the *Secondary* period, with highly stratified rocks, sandstones, limestones, gypsum and many other layers, filled with fossils. Finally, the most recent were the generally unconsolidated deposits of gravels, sand and clays, corresponding to the *Tertiary* period.

Rather than accepting that the earth's crust had formed in a mere six thousand years, Abraham Werner speculated that the older Primary and Transition rocks had formed more than a million years ago, by precipitation from a universal ocean that once enveloped the whole world. His theory implied that the order of rocks he had identified in Saxony would be found elsewhere. If his observations were right, the consequences of his findings were huge, as they were proof that locked within the earth's crust was evidence of distinct periods in its formation. By identifying an order in the layers of rock, Werner was offering the world a glimpse of prehistory.

Even more perplexing amid the lecture-rooms of deans and bishops at Oxford was a new theory put forward by a Scotsman, James Hutton. He did not accept Werner's view that the older rocks had precipitated from a universal ocean, but envisaged that they were formed gradually by erosion and deposition. This led him to speculate that the history of the earth was so vast it was almost immeasurable.

From his observations Hutton inferred that the earth was caught in an endless cycle of forming and re-forming the landscape: cycles in which rivers carried sediment from the land to the sea; layers of sediment gently accumulated and compacted into stone on the sea floor, until earth movements lifted the layers out of the sea, folding the different strata to form a new landscape. Since, he reasoned, the erosion of land and the accumulation of sediment take millions of years, the only conclusion was that the landscape had formed over millennia. In his book,

Theory of the Earth, he wrote of the earth's history, 'there is no vestige of a beginning and no prospect of an end'.

The ideas of Hutton, Werner and others opened the door to an unfamiliar landscape as well as a vast, unknown history of the earth. This abyss of geological time was almost as strangely unbelievable as the vastness of stellar space opened up by Copernicus in astronomy two centuries earlier. The new theories questioned the long-established chronology for the earth's age of Archbishop Ussher, and with it, the authority of the Bible. Many thinkers felt that this was a dangerous pursuit. Richard Kirwan, President of the Royal Academy in Ireland, was one of several leading thinkers to ridicule Hutton's claims, pointing out that this was 'fatal' to the account in Moses and therefore a threat to morality. Hutton's theory was so obviously flawed that Kirwan had found it quite unnecessary to even read it!

William Buckland, brought up in the heart of the Anglican establishment but drawn to a rigorous, scientific approach to gathering evidence, was eager to understand the true history of the globe within which fossils could be correctly understood. Wishing to reconcile the two seemingly opposing sides of his nature, he dedicated himself to proving that religion and science did not stand opposed to each other, but were complementary. For him, geology was a 'master science' through which he could investigate the signature of God.

In 1813, when Buckland was appointed Reader in Mineralogy at Oxford, such was his enthusiasm to make sense of the apparently conflicting opinions about the earth's history that he embarked on a detailed study of all the rocks of England, travelling with his friend, George Bellas Greenough. Greenough had helped to found the Geological Society of London in 1807. This began as a 'little talking Geological Dinner Club' in a central London tavern, and had rapidly blossomed into a scientific society which aimed to 'make geologists acquainted with each other, stimulate their zeal . . . and contribute to the advance of Geological Science'.

Touring with Greenough, Buckland aimed to construct a geological map for the Society of all the strata they could identify, showing the

The Reverend William Buckland, 1832.

different layers of rock in each region and comparing the fossils within them. Would the layers of rock in England correspond to those in the European rocks? What did the different formations reveal about pre-history?

With infectious enthusiasm, Buckland also enlisted the support of his long-standing friend, the intellectual Reverend William Conybeare, who had graduated from Oxford just before Buckland, taking a first in classics at Christ Church with effortless ease. The unconventional party also called upon 'the zealous interest of some ladies of high culture at Penrice Castle, Lady Mary Cole and the Misses Talbots', and any other like-minded individuals they met along the way. Buckland's energetic and novel approach, which would not be constrained by centuries of Oxford tradition, was viewed with more than a little suspicion.

Whereas most gentlemen's travelling carriages would have been of a certain standard, with an elegantly appointed interior matched, perhaps, by a smartly painted exterior and with discreet uniformed staff in attendance, William Buckland's carriage provided a very different travelling experience. The sturdy frame was specially strengthened to allow for heavy loads of rocks; the front was fitted with a furnace and implements for assays and analysis of the mineral content of the stone; and there was scarcely room to sit amid the curiosities and fossils heaped into every available space.

Gossip abounded, too, about Buckland's other little eccentricities. It was the custom for early geologists to carry out their fieldwork in the full splendour of a gentleman's suit, with academic robes and even a top-hat. When travelling in the mail, Buckland was not beyond dropping his hat and handkerchief in the road to stop the coach if he spotted an interesting rock. On one occasion he happened to fall asleep on the top of the coach. An old woman, eyeing his bulging pockets with growing interest, eventually couldn't resist emptying them, only to find to her astonishment that the gentleman, for all his finery, had pockets full of nothing but stones.

Sometimes Buckland rode a favourite old black mare, usually burdened with heavy bags and hammers. It was said that the mare was so accustomed to her master's ways that even if a stranger was riding her, she would stop at every quarry and nothing would persuade her to advance until her rider had dismounted and pretended to examine the surrounding stones. Buckland became so expert on the rocks of England that his 'geological nose' could even tell him his precise locality. Once, when riding to London with a colleague on a very dark night, they lost their way. To his friend's astonishment Buckland dismounted and, grabbing a handful of soil, smelled it and declared, 'Ah Uxbridge!'

William Conybeare, it seems, was as zealous in his search for fossils as Buckland, and their activities never failed to attract attention. Once on a tour together they entered an inn after a particularly long, wet day on the cliffs, covered in mud and dirt. The two deans had fossil bags filled to bursting and proceeded to empty out the contents. The old woman serving their meal was said to be 'much puzzled to make out the Deans' real character'. After eyeing her ravenous customers suspiciously, she exclaimed, 'Well I never. Fancy two *real* gentlemen picking up stones! What won't men do for money!'

In trying to create a map showing the order of succession of the rock strata of England, Buckland and his friends were greatly influenced by the pioneering work of a surveyor called William Smith. A man of humble birth, Smith lived at the height of the 'Canal Age' in the late eighteenth century, when the fields of England were criss-crossed by a network of over two thousand miles of inland waterways. As he surveyed the land for canal building, he had become very familiar with the sequential order of British rock from the chalk down to the coal. He noticed that different strata contained different fossils and that this could be used to help identify some of the layers. Such was his enthusiasm to understand the order of strata that Smith devoted his modest income to travelling all over England. Versions of his geological tables had been on display since the 1790s, and he published his great map *A Delineation of the Strata of England and Wales* in 1815.

Unfortunately for Smith, George Bellas Greenough, the first President of the Geological Society, had little time for him and his map. When he saw Smith's tables he was condescending and patronising and yet, it has been argued, with 'barefaced piracy' he was able to draw heavily on this work for the benefit of the Society. Undoubtedly Smith's studies laid the groundwork for Buckland, who between 1814 and 1821 produced no less than eight different charts of the 'Order of super-imposition of strata in the British Islands'.

All of this made little impression on the canons and bishops at Oxford. Scholars and religious leaders were alarmed that the sacred evidence of the word of God should be muddied with bits of rock and dirt. 'Was ever the Word of God, laid so deplorably prostrate at the feet of an infant and precocious science!' exclaimed George Bugg, author of *Scriptural Geology*. 'We want no better guide than Moses,' wrote George Cumberland to the editor of the popular *Monthly Magazine* in 1815. 'If the object of geology be to attain the age of the earth as a planet, it seems an idle proceeding; first because if attained, it would apparently be useless . . . it can never be attained by the present mode of enquiry; and like the riddle of the Sphinx, would destroy the life of those who failed in solving it, by wearing out the only valuable property they have, viz, their intellects!'

For years, dons wielding authority through their sermons and sacred texts had successfully kept alternative schools of thought at bay. Among the more traditional scholars there was a real fear that geology would prove to be a 'dangerous innovation', and Buckland's odd activities were watched 'with an interest not wholly devoid of fear'. At the end of the Napoleonic Wars in 1816, when Buckland took the opportunity to travel with Conybeare and Greenough across Europe, his departure was welcomed by some of the elderly classicists at Oxford. 'Well Buckland has gone,' announced one dean with satisfaction. 'Thank God we will hear no more of this Geology!' Nothing could have been further from the truth.

In 1816, Buckland published the first comparative table of the strata of England compared with those of the Continent. Similarities between the rocks of England and Europe were beginning to emerge. Greywacke slates, resembling the continental Transition formations, were found on the borders of England and Wales. Highly stratified layers of sand-stone, limestone and conglomerates rich with fossils, like the Secondary formations of Europe, were widespread across England. Tertiary rocks, such as those around Paris, were identified in the London and Hampshire basins. Just as in Europe, these were always in the same order of succession, the oldest being Primary, then Transition, Secondary and Tertiary. As correlations were found between different regions, 'marker' rocks were identified. Chalk, for instance, was recognised as the upper limit of Secondary rock throughout Europe.

Buckland was keen to discover whether this order of succession extended worldwide. He wrote to several noblemen in command of Britain's growing Empire, such as Lord Bathurst, the Secretary of the British Colonies, enclosing instructions for collecting geological specimens abroad. His appetite for information became insatiable: it was as if the layers of rock that enveloped the globe formed the pages of a history of the earth. But if this was so, what would be written on them? And how did all this fit with the extraordinary 'crocodile' found by Mary Anning?

The first clue to this puzzle lay in a remarkable new approach to interpreting fossils that was being pioneered in Paris by a French naturalist called Georges Cuvier. From a poor but bourgeois family, Cuvier had survived the French Revolution in Normandy, far from the troubles of Paris, where in his letters he had feigned support for the regime for fear of the French police. Once the Reign of Terror had released its grip on Paris and the city became safe again, Cuvier went to the capital and soon secured a post at the Muséum National d'Histoire Naturelle. With his striking crop of red hair, bright-blue eyes and somewhat unkempt appearance, it wasn't long before the ambitious young naturalist had made an impression.

As Napoleon's army swept across Europe, spoils from museums and private collections were frequently sent back to Paris. Fossils were also retrieved from the plaster quarries around Paris, and during the course of building canals around the city. The new Muséum National d'Histoire Naturelle, established by the Republicans in place of the Jardin du Roi, rapidly became the envy of the world. Cuvier began to apply his extensive knowledge of the anatomy of living creatures to try to interpret fossil skeletons with a view to understanding the ancient forms of life.

Georges Cuvier believed that fundamental laws must govern the anatomy of creatures as surely as the laws established by Newton now governed physics. If a creature was a carnivore, Cuvier observed, all of its organs would be designed for this purpose. The forelimbs would be strong enough to grasp prey; the hind-limbs muscular and mobile, for hunting; the teeth would be sharp, capable of ripping meat; the jaw would have sufficient muscular support for the animal to tear prey; and the digestive organs would be adapted for carnivorous food. In effect, Cuvier's principle of 'correlation of parts' showed that all the organs and limbs of a creature are interdependent and must function together for that creature to survive. He rapidly acquired a brilliant reputation. From a single fossil bone, he declared, he could deduce the *class* of the beast – whether it was a mammal, reptile or bird – and ascertain subordinate divisions: the *order*, *family*, *genus* (plural: *genera*), and perhaps even the very *species* to which the fossil animal belonged.

'Let us not search further for the mythological animals,' said Cuvier. 'The mantichore or destroyer of men which carries a human head on a lion's body terminating in a scorpion's tail, or the guardian of treasures, the Griffin, half eagle–half lion . . . Nature could not combine such impossible features.' The teeth and jaws of a lion, for example, could only belong to a creature that possessed the other attributes of a powerful carnivore, a muscular frame and skeleton that would confer enormous strength. The Sphinx of Thebes, the Pegasus of Thessaly, the Minotaur of Crete, mermaids – those half-women half-fish that lured sailors to their death with the sweetness of their song – were all myths that crumbled under Cuvier's scientific scrutiny. 'These fantastic compositions may be recovered among ruins,' he said, 'but they certainly do not represent real beings.' Instead, Georges Cuvier offered a real past, conjuring up a vivid picture of creatures that had once roamed the surface of the earth.

Less than two years after his arrival in Paris, in January 1796, the twenty-seven-year-old naturalist made his debut at the National Institute of Sciences and Arts. His talk 'On the species of living and fossil elephants' pointed to an astonishing conclusion.

Following French victories in Holland, a private collection of fossil 'elephants' at the Hague had been seized and sent to Paris. Cuvier had compared these fossils from Holland to the bones of present-day elephants from India and Africa. As he studied the characteristics of the teeth and jaw he realised that the fossil 'elephant' differed in the shape and proportions of the jaw from either of the two living species. On the basis of these differences, he argued, the fossil 'elephant' should be classified as a separate species. The distribution of the fossil bones also differed; unlike the Indian or African elephant, the fossil species was never found in the tropics. He gave the fossil elephant a special name in recognition of its differences: the 'mammoth'.

Since mammoths differed from any living elephants, reasoned Cuvier, this species was now extinct. The discovery, soon after this, of the first preserved mammoth in the permafrost of Siberia lent weight to his ideas. Cuvier believed the snowy wastes of Northern Europe and Siberia had once been inhabited by these enormous woolly beasts, which had some-how mysteriously perished. And he went on to show that other large fossil mammals, apart from the mammoth, had thrived on the ancient globe. He identified '*Megatherium*', or 'huge beast', a creature resembling a giant sloth and covered in fur like a bear, which could stand on two legs to graze on leaves. An elephantine creature whose fossils combined the teeth of a hippopotamus with the huge tusks of a mammoth was named by Cuvier a 'mastodon'.

Cuvier's large extinct mammals, the mammoth, the mastodon and *Megatherium*, were found in the most recent, Tertiary deposits. In older strata Cuvier identified an ancient sea lizard, '*Mosasaurus*' or 'lizard of the Meuse', and several extinct species of crocodile. His studies suggested that entire animal races had been wiped from the face of the earth. He was haunted by the desire to know what had happened to the vanished creatures. Why would God create these beings if He planned only to destroy them? Cuvier wanted to ascertain whether 'species which existed then have been entirely destroyed, or if they have merely been modified in their form, or if they have simply been transported from one climate into another'. Quite why and how extinction occurred was a puzzle that remained to be solved.

William Buckland was impressed by Cuvier's discoveries and eager to learn from his approach, comparing fossil animals to living creatures so as to work out their zoological affinities. He discussed Mary Anning's unknown creature with his friend the Reverend Conybeare, who wanted to make a definitive scientific study of the giant beast. Mary's 'crocodile' possessed such a puzzling blend of characteristics that it was hard to classify. The long, pointed snout was similar to a dolphin's or porpoise's. The teeth were more like those of a crocodile, with sharp, conical fangs, each one ridged all around the enamel. The vertebrae were slender, like the backbone of a fish. It was baffling.

To compound their problems, England did not have a centre of anatomical excellence comparable to the magnificent collections under Cuvier's supervision in Paris. Consequently, Buckland tried to establish a correspondence with Cuvier, 'founded on an exchange of fossil specimens', and hoped to benefit from the French expertise.

It was to Lyme that the Reverends Buckland and Conybeare went in search of fossil 'crocodiles' as gifts for Cuvier, and in particular to the collection of Mary Anning.

Mary and her mother had established a 'tiny, old curiosity shop close to the beach'. According to one visitor, 'the most remarkable petrifactions and fossil remains . . . were exhibited in the window'. Inside, the little shop and adjoining chamber were 'crammed with ammonites, heads of "crocodiles", and boxes of shells'. To Mary's skills as a collector, Buckland acknowledged, he felt greatly indebted, for she continued to supply more specimens of her unknown creature. Cuvier was interested to see the latest discoveries from England, and soon Buckland established a correspondence with a young assistant in Cuvier's department, Joseph Pentland. Pentland acted as liaison between Cuvier and the English team, organising shipments of casts and providing information on fossils.

But while Buckland and his colleagues were approaching Georges Cuvier, another London gentleman, Sir Everard Home, raced into print with the first published account of Mary's creature. Although Sir Everard relished his reputation as Britain's leading anatomist and held the distinguished position of Surgeon to the King, he was in fact not only incompetent, but also a fraud. Much of his fame was due to reflected glory from John Hunter, his famous brother-in-law.

John Hunter was revered in England as the 'father of modern surgery' and had pioneered early studies of anatomy before his sudden death from a heart attack. Sir Everard was secretly plagiarising Hunter's unpublished manuscripts. He had removed 'a cartload' of Hunter's anatomical papers from the Royal College of Surgeons in London. Once he had copied them out in his own name, he allegedly burned Hunter's originals. Such was his enthusiasm to demolish the evidence, on one occasion Sir Everard set fire to his own hearth and had to call out the fire brigade.

In his first paper to the Royal Society in 1814, Sir Everard initially favoured the idea that Mary Anning's creature was some kind of crocodile. This was because he had noticed small germs of conical teeth contained within the larger teeth. Whereas mammals have just two sets of teeth, the milk teeth and the adult teeth, reptiles have replacement teeth growing through the jaw all their lives. But when Sir Everard split one of the teeth open, he mistook the young germ tooth inside for an accumulation of calcareous minerals. 'The characteristic mark therefore, of a crocodile's teeth,' he wrote, 'was thus removed.' He wrongly concluded that it was not a reptile.

Then he reasoned that it must be an enormous aquatic bird, since the pattern of openings in the skull of the creature was similar to that of birds. The bones of the eye, he wrote, 'subdivided into thirteen plates, which is only met in birds'. But if it was a bird, where were the wings, and why so many fish-like characteristics? Sir Everard considered that the lower jaw of the skull 'admits the mouth to be opened to a great extent . . . resembling the voracious fishes'. New specimens revealed the 'bird' had paddles for swimming, and he decided the creature belonged to the class of fishes; although, somewhat baffled, he wrote, 'I by no means consider it wholly a fish.'

After his initial uncertainty over whether the beast should be classed as reptile, bird or fish, by 1819 Sir Everard thought he had solved the puzzle. A new creature called a 'Proteus' had just been described in English by a Viennese physician. This was a blind, amphibious, serpentine creature with very unusual anatomical features that inhabited caves. Mistakenly guessing that the Lyme 'crocodile' was a link between the Proteus and lizards, he named it 'Proteosaurus', or 'Proteus-lizard'. However, the year before, Mary Anning's creature had been sold to the British Museum, where the Keeper of Natural History, Charles Konig, had named her animal '*Ichthyosaurus*', meaning 'fish-lizard'. This was in recognition of its curious mixture of fish and reptile characteristics. Since this name had been put forward first, it had priority over any other. Sir Everard Home was furious, and he continued to promote his own rival name, 'Proteosaurus'.

In all this confusion, one thing was clear: the French were laughing at the English grasp of anatomy. Joseph Pentland, in Cuvier's laboratory, scoffed at the papers of the 'London Baronet', as he called Sir Everard. He wrote to William Buckland in Oxford saying that Sir Everard's 'ridiculous' papers were 'abstruse, incomprehensible and for the most part, uninteresting'. What is more, the London Baronet was 'crowding' the *Philosophical Transactions of the Royal Society*, the prestigious journal of the oldest scientific society of Europe, blocking the publications of others whose work was more 'worthy and honourable'.

Possibly because Sir Everard dominated the Royal Society, Buckland's friends, the Reverend Conybeare and another enthusiastic young geologist, Henry de la Beche, prepared their detailed scientific paper on Mary's creature for the Geological Society. They gathered many more specimens from Lyme and the Bristol area and were also able to capitalise on the anatomical expertise of the French. 'I am sure that the fossil approaches much nearer to the family of Saurians [lizards],' wrote Pentland to Buckland in 1820. 'The dentition of the Ichthyosaurus is the same as in lizards.'

Conybeare and de la Beche published their findings in 1821. In agreement with the French, they showed that the teeth of the animal bore more resemblance to those of a crocodile than to any other creature. The replacement cycle of teeth so characteristic of a reptile, with 'the young tooth growing up in the interior cavity of the old one,' wrote Conybeare, 'is *exactly* similar'. The bones of the skull were also lizard-like, with two openings at the back behind the eye, lightening the skull and allowing the muscles of the jaw to bulge so that it could work more efficiently. In the lower jaw alone, all the bones that Cuvier had identified in a crocodile could also be seen in this animal.

There were, however, some differences between Mary's fossil and a crocodile skull. The teeth, Conybeare observed, 'are more numerous than in the crocodile, there cannot be less than 30 a side'. The huge round eyes were larger in proportion to the skull than the eyes of any other known animal. Having no eyelids, to prevent injury in a rough sea, it had instead many thin, flexible bones encasing the pupil to protect it. The general shape of the jaw, he thought, 'differs from the crocodile in being much more lengthened', and ending in a point 'almost as sharp as the beak of a bird'. Nonetheless, in both the dentition and the bone structure the animal 'approaches more closely to the Saurian or Lizard family, and especially to the genus Crocodile,' said Conybeare, 'than to any other recent type'. The fossil beast, therefore, belonged to the reptile class and the saurian family.

Despite this, it had many characteristics of fishes. The vertebrae were

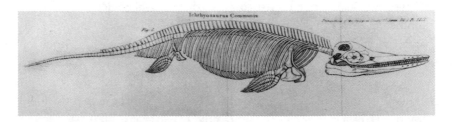

The skeleton of the *Ichthyosaurus*, up to 17 feet long,
found by Mary Anning in 1812.

just like those of a fish, with small, flat discs allowing enormous flexibility of the spine. The bones were also very light, combining the 'greatest strength with least weight', which would 'increase the buoyancy of the animal and enable it to face the waves of an agitated ocean'. With eighty or ninety such vertebrae, the creature could reach twenty-four feet in length. In view of its fish and lizard affinities Conybeare accepted the name *Ichthyosaurus*, or 'fish-lizard', to denote the genus. While tactfully acknowledging the 'praise worthy readiness' with which Sir Everard had communicated his ideas 'instantly to the public', his 'Proteosaurus' was quietly forgotten. *Ichthyosaurus*, said Conybeare, roamed the primitive seas 'upon which no human eye ever rested'. He tried to trace the boundaries of this long-buried sea by seeing how far the fossil remains extended across England. They found ichthyosaurs in many counties in South-west England deposited within the Secondary strata.

As Conybeare and de la Beche searched the Secondary rocks, they came upon other bones, principally vertebrae, which did not quite match those of *Ichthyosaurus* or of a crocodile. 'I was persuaded that they had all belonged to different places in the vertebral column of a single species,' wrote Conybeare. He began to suspect that another unknown sea lizard had shared the ancient ocean with the ichthyosaurs. He proposed the name 'Enalo-sauri', or 'sea lizards', to denote the whole order, and hinted strongly that more types of these giant sea creatures had yet to be uncovered. The paper was seen as a triumph, and their description of the ichthyosaurs stands to this day.

As for Mary Anning, she hadn't the education or the position in the world to name her finds or to use them as an entrée to the male-dominated world of science. She was not even named in the scholarly papers on her creature published in London. In her cottage by the sea or sitting on the shore at Lyme, she painstakingly copied out the learned articles in her own hand, making drawings and trying to grasp the language of the new science. There is even a suggestion that she may have tried to learn French in order to read Cuvier for herself. With many French visitors to the port of Lyme, this was not such an impossible feat.

Mary was sufficiently encouraged by her first discovery to persevere in her daily searches on the shore, braving all weathers. The deplorable conditions of five years' parish relief focused her efforts tremendously as, according to one collector, Thomas Hawkins, she 'explored the frowning and precipitous cliffs, when the furious spring-tide conspired with the howling tempest to overthrow them, and rescued [fossils] from the gaping ocean, sometimes at the peril of her life'. The dangers Mary faced were also noted by a gentleman's daughter, Anna Maria Pinney, who sometimes explored the cliffs with her: 'we climbed down places, which I would have thought impossible to have descended had I been alone. The wind was high, the ground slippery, and the waves beating against Church Cliff. When we had clambered to the bottom our dangers were by no means over . . . In one place she had to make haste to pass between the dashing of two waves . . . she caught me with one arm round the waist and carried me some distance.'

As news of Mary Anning's finds spread among the members of the Geological Society several gentlemen, as well as William Buckland, sought her out at Lyme. She was cultivated by Henry de la Beche, who was studying the *Ichthyosaurus* with Reverend Conybeare. De la Beche was a young man of independent means who had inherited from his father an estate in Jamaica, which had prospered with the slavery trade. A Lieutenant-Colonel Thomas Birch also took a keen interest in gathering fossil evidence of the *Ichthyosaurus*, and acquired many of her specimens. Anna Pinney noted that Mary was 'courted by those above her', and she rapidly acquired 'many ideas and a power of communicating them'. In spending time with such gentlemen from a very different class, she had already stepped aside from her peasant background. 'She frankly owns,' admitted Anna, 'that the society of her own rank is become distasteful to her.' Despite this, she continued to 'attend the sick poor night and day, even when they are ill with infectious diseases'. Whether Mary dared to hope that one day she might escape hardships of her upbringing through marriage is not recorded.

She became a familiar figure on the shoreline, variously portrayed in her long skirts and shawl, clogs, poke-bonnet or hat, a lone figure endlessly toiling at her mysterious task against vast skies and shifting

Mary Anning, 1841.

tides. Such was her dedication, Anna Pinney wrote, that she continued 'to support her mother and brother in bitter poverty even when she was so ill that she was brought . . . fainting from the beach'.

The layers of rock that so fascinated Mary Anning held the secrets of prehistory. Locked behind the impenetrable dark face of Black Ven and the cliffs beyond were the clues to an ancient ocean, whose boundaries were yet unknown. From her discussions with the gentlemen geologists, Mary knew that another kind of sea lizard was almost certainly buried there, waiting to be uncovered.

2

The World in a Pebble

There is no picking up a pebble by the brook side without finding
all nature in connexion with it.

Cited in *Thoughts on a Pebble*
by Gideon Mantell, 1849

While Mary Anning was searching the shore for fossils, a young
shoemaker's son, Gideon Algernon Mantell, was trying to make
his own way in the world of science. A story told by one of his childhood
companions reveals that, like Buckland, Gideon Mantell was drawn to
geology early in life:

> As a mere youth, he was walking with a friend on the banks of
> the River Ouse when his observant eye rested on an object
> which had rolled down the marly bank . . . He dragged it from
> the water and examined it with great attention. 'What is it?'
> inquired his friend. 'I think that it is what they call "a fossil",' he
> replied. 'I have seen something like it in an old volume of the
> *Gentleman's Magazine*. The curiosity, which proved to be a fine
> specimen of Ammonite, was borne home in triumph . . . and
> from that moment young Mantell became a geologist.

It was a revelation to Mantell that buried in the earth beneath their
feet lay the 'wreckage of former lives that had turned to stone'. His
home town of Lewes in Sussex is enveloped by the dramatic contours of

An ammonite: according to ancient folklore, these
extinct marine invertebrates were thought to have
magical powers.

the chalk South Downs. Past the grammar school, the castle and the
Market House, the High Street plunged towards the valley of the River
Ouse and the chalk spur beyond loomed above the smoke from the chim-
neys of the shops. To the south, past the ruined priory, the green fields,
decked with white wherever the chalk broke through the thin covering
of grass, beckoned Mantell through every cobbled alley-way.

As a child exploring the local pits and quarries he uncovered
ammonites with their coils 'like the fabled horn of Jupiter, Ammon', and
shells with spines, such as the sea urchin, and the remains of corals
and fishes; the chalk hills teemed with the worn relics of creatures that
had lived long ago. For the young Mantell, science was 'like the fabled
wand of the magician' which could 'call forth from the stone and from
the rock their hidden lore and reveal the secrets they have so long
enshrined'. Every fossil reclaimed from the past was, for him, a 'medal
of creation', a fantastic page of Nature's volume to interpret.

Far removed from his vision of ancient worlds was the daily reality of
supplying the town's footwear. He and his six brothers and sisters were

brought up in a cottage in St Mary's Lane, a steep, narrow road that ran off the High Street. At a time when social status was principally determined by money and land, Gideon Mantell was aware of his family's modest station in life. Although his father, Thomas, ran a successful business, sometimes employing several people, he was a 'tradesman', not a 'gentleman', and so excluded from the higher ranks of society.

It was a far cry from what Mantell understood of the great wealth of the family's forebears. In his youth, he dreamed of restoring the family honours. He told a friend, 'although my parents and their immediate predecessors were in comparatively humble stations, being only trades people in a country town, yet they were descendants of one of the most ancient families in England. The name "Mantell" occurs in the list of Knights that accompanied William the Conqueror from Normandy. The family settled in Northamptonshire and possessed large manors at Heyford and Rode where many of the family bore the honor of a Knighthood.'

But the family fortunes had been lost almost overnight. The grandson of Sir Walter Mantell, a Protestant, took part in Sir Thomas Wyatt's attempt in 1554 to prevent the Catholic marriage of Queen Mary with Philip of Spain. The planned royal marriage was so unpopular that Wyatt and an array of four thousand men almost reached London Bridge before they were outnumbered and eventually forced to surrender. Wyatt and the ringleaders, including Mantell and his grandson, were executed. As if this was not enough, all the Mantell family estates, in Kent, Sussex and Northamptonshire, were forfeited to the Crown. 'Irretrievable ruin fell upon the house,' wrote Gideon Mantell; 'in my boyish days I fancied I should restore its honors and that my children would have obtained the distinctions our knightly race once bore.'

Mantell's soaring ambitions were not without foundation, for he was regarded as something of a child prodigy in his home town. He was distinguished by 'uncommon perseverance and quickness in his studies'. Owing to his pious parents, 'his retentive memory enabled him when young to repeat a large part of the Bible by heart'. When older, he was described in local records as 'tall and graceful', and with a 'style of brilliancy and eloquence'. A painting of him in his youth shows a handsome face, with even, expressive features and dark hair and eyes. Whether this is a truthful portrait is unknown, but according to the *Sussex Gazette* he was not lacking in charisma: 'He had the attractive personality of an actor, a voice of great power, and with clear enunciation and pleasing musical cadences he could hold his listeners spellbound.'

But as the son of a bootmaker, the young Gideon Mantell was educated with great frugality. Because of his father's nonconformist beliefs as a Methodist, the six children were excluded from the local grammar school; the twelve free places each year were reserved for those brought up in the Anglican faith. Instead, Gideon was sent to the dame-school among the labourers' cottages in St Mary's Lane. Here, under the simple guidance of an old woman, he was taught the rudiments of reading and writing in her front parlour, and he became so great a favourite that on her death the teacher left Gideon everything she had. After this, he went to the school of a Mr John Button, an exuberant philosophical radical, 'where a sound and practical commercial education was given by a gentleman whose political sentiments were so accordant with those of Gideon Mantell's father, that he was known to be on the Government black list'.

Mr Mantell's political views are not stated; however, as a radical Whig, it seems likely that he associated with the campaigning Thomas Paine, well known reformer and also an inhabitant of Lewes. Paine was a keen debater at the Headstrong Club which met at the White Hart in the High Street. He openly challenged the value of the British monarchy at a time when the Revolution raged in France, he denounced cruelty to the poor, demanded the abolition of the slave trade, and later wrote *The Rights of Man*.

After two years with Mr Button, Gideon was sent away for a period of private study with his uncle, a Baptist minister, who had founded a 'Dissenting Academy for Boys' near Swindon. When he returned to Lewes at the age of fifteen, with the assistance of the leader of the local Whig party who was impressed by his diligence Gideon was apprenticed

Gideon Mantell, 1837.

to a local surgeon, James Moore. On his father's death in 1807 money was found for him, in the last year of his 'bondage' as apprentice, to study in London and 'walk the hospitals'.

At seventeen, Mantell went to London to study medicine, carrying a bag full of fossils collected from the chalk hills of Sussex. These curios, somewhat unnecessary for a student doctor, were nonetheless of such importance to Mantell that he had found room for his 'extensive collection' on the stagecoach to London. But if he was hoping for an opportunity that would allow him to immediately develop a career in geology he was soon to be disappointed. There were, as yet, no academic posts in the subject, and his father's Methodism and his educational background precluded him from university.

The main forum for geologists was in the scientific societies springing up in the metropolis such as the Geological Society and the longer-established Royal Society. But they were largely for gentlemen of rank and wealth, and gaining membership cost time and money. Of these, the Royal Society was the most famous; its Council had provided instructions for Captain Cook's voyage of discovery and advised the government on scientific matters such as the best form of lightning conductors for buildings. The membership list read like the entries of the fashionable new guide to Society, Debrett. Lords, knights and men who 'from their fortunes it might be desirable to retain as patrons of science' dominated the list of Fellows. A shoemaker's son, however brilliant, was largely invisible to this scientific community. But while in London, a chance meeting was to set Gideon Mantell on his future course.

In 1811, the year Mary Anning's brother found the skull of the *Ichthyosaurus*, a distinguished doctor, James Parkinson, published the final volume of his studies on geology, *Organic Remains of a Former World*. Mantell may have been drawn to Parkinson because, like his father, he was a man of conscience, interested in reform. Parkinson had published such inflammatory pamphlets as *While the Honest Poor are wanting Bread* and *Revolution without Bloodshed*, advocating universal suffrage. He had even come dangerously close to transportation to Australia in 1794, when he was arrested for an alleged connection with the 'Pop Gun Plot' to assassinate King George III with a poisoned dart while he was at the opera. He was exonerated from treason, but after this incident he restricted his political interests to social reforms, to improving conditions for pauper children, and to treatments for the insane in asylums. James Parkinson is now better remembered as the doctor who first defined 'Parkinson's disease', the degenerative illness marked by shaking and tremors.

At the beginning of the nineteenth century, though, Parkinson was equally well known as a geologist. Along with William Buckland and George Greenough, he was one of the founder members of the Geological Society and had embarked on a detailed survey of everything known about the 'Ante-Diluvian World'. To Mantell in the Lewes library eagerly taking in the descriptions of the entire vegetable and animal fossil kingdom, Parkinson's work was an inspiration. Putting aside any scruples about imposing on such an eminent gentleman, he made an appointment to visit Parkinson in Hoxton Square, Shoreditch, in East London.

His nervousness at seeking 'the pleasure and the privilege' of such an acquaintance was soon dispelled by James Parkinson's 'mild, courteous manner', Mantell wrote, and the enthusiasm with which he 'explained to me the principal objects in his cabinets and pointed out every source of information on fossil remains'. Parkinson had assiduously gathered details of Georges Cuvier's studies in Paris and could tell Mantell of his famous discoveries: the giant extinct mammals, the mastodon, *Megatherium* and mammoth, and ancient species of crocodiles found around Honfleur and Le Havre. Cuvier believed that the fossil bones of crocodiles came from limestone beds of 'very high antiquity . . . considerably older than those which contain the bones of quadrupeds'.

Parkinson had been greatly influenced by the pioneering work of the surveyor William Smith. Whereas Werner, in Saxony, identified rocks principally on the basis of their mineral composition, Smith had recognised that fossils could be used to help identify the beds. In his publication of 1811 Parkinson was careful to classify fossils according to the strata in which they were found; each layer of rock with its entombed fossils was for him a 'former world' which held the secrets of the history of the globe.

Parkinson, like Buckland, was intrigued by the conflict between geology and religion and was resolved 'to shrink from no question . . . however repugnant to popular opinion'. He concluded that the account of Moses in the Bible 'is confirmed in every respect, except as to the age of the world, and the distance of time between the completion of different parts of Creation'. Although there was no way of proving the earth's antiquity, he acknowledged that the formation of the globe and the creation of life 'must have been the work of a vast length of time'. Following an idea first raised by scholars in the eighteenth century, he reasoned that if the word 'day' in Genesis was used 'to designate *indefinite periods* in which particular parts of the great work of Creation was accomplished, no difficulty will then remain'.

Parkinson fired the young Mantell with his romantic description of 'former worlds' buried in the rock. Each stratum enveloped evidence of a vanished existence, and the geologist could 'begin to fathom the different revolutions which had swept over the earth in ages antecedent to all human record or tradition'. Parkinson wrote: 'even the enormous chains of mountains which seem to load the surface of the earth are vast monuments in which these remains of former ages are entombed . . . they are hourly suffering those changes by which after thousands of years they become the chief constituent parts of gems; the limestone which forms the humble cottage of the peasant, or the marble which adorns the splendid palace of the Prince.' The mountains, the hills and the land beneath their feet: all these were vast tombs more astounding than the pyramids.

It was through meetings with men like Parkinson that Mantell's ambitions began to take shape. It was, he thought, the role of the scientist 'to unveil God's secrets . . . and unravel the mysteries of the beautiful world through which he was destined to pass'. James Parkinson had found time for geology while practising as a doctor. Mantell, too, would carry on his childhood dream. He would devote every spare minute to exploring these ancient memorials to a buried past that had existed, it seemed, before Adam. When he returned home to Sussex, he planned to make a systematic study of the strata and fossils of the county, a subject which he viewed as 'replete with interest and instruction'. This married together his fascination with the subject and his desire to make a name for himself that might bring back honour to his family name.

At the age of twenty-one he gained his diploma of membership of the Royal College of Surgeons and returned to Lewes, where he was immediately offered a partnership with his former master, James Moore. It was soon apparent that he faced a gruelling workload as a country doctor. Epidemics of cholera, typhoid and smallpox still raged. 'An immense number of Persons in this Town and neighbourhood are ill with Typhus Fever,' he recorded in his diary on one occasion. 'I have visited upwards of 40 or 50 patients every day for some time: yesterday I visited 64. The small pox is also very prevalent, 14 have died with it, Taylor in Malling Street who had it in 1794 is now covered with various pustules and has been very ill.' Armed with little more than boxes of leeches, which were sent from London in boxes of two hundred, he struggled on against these deadly diseases.

His practice included attendance on the sick poor of three parishes, for which he was paid £20 a year, and treating the inmates of the Poor House at St John's, near Lewes. Long before the development of emergency services, the local doctor provided the only care, even for the severely injured: 'This morning I was summoned to Ringmer, a poor woman on the Green, Mrs Tasker, had set fire to her clothes and was most dreadfully burnt, it is scarcely possible she should survive.' For five weeks he visited for an hour every day to change her dressings. When she died, he noted, 'I am almost fatigued to death.'

There were numerous mills in the district for grinding corn, for producing rapeseed oil and flour, and for brewing or malting, and even for producing paper. With child labour, accidents were common and, without anaesthetic, invariably traumatic: 'I was called to a most distressing accident at Chailey Mill. A poor boy got his clothes entangled by an upright post that was rapidly revolving; the lad in consequence was whirled round with great velocity and his legs were dashed against a beam. It was considered absolutely necessary to amputate the left leg above the knee, but the constitutional shock was so great that the poor boy died the next day.'

The hours of Gideon Mantell's medical practice were long and unpredictable, especially since he excelled at midwifery, delivering between two and three hundred babies a year. At a time when the average mortality for women in some hospitals was as high as one in thirty, Mantell had only two deaths in over two thousand births during fifteen years. His great success came at a price: 'frequently I have been up for six or seven nights in succession: an occasional hour's sleep in my clothes being the only repose I could obtain'. Nonetheless, with his conscientiousness and tireless energy he gradually increased the profits on the practice from £250 a year to £750.

Despite the immense pressures of his medical workload, Mantell was prepared to sacrifice his few leisure hours to make headway in geology as well. With the carelessness of youth he spared himself nothing, often studying until the small hours and rising after just four hours' sleep, to see his patients before embarking on some geological expedition.

There were numerous local pits and quarries such as Jenner's quarry, Malling Hill Pit, Malling Street Pit, Southerham Pit, each laying bare the strata of Sussex. Following the approach of William Smith, he aimed to construct a geological chart of the correct sequence of the local rocks. He paid the pit labourers for any interesting fossils that he could add to his cabinet, and soon became familiar with the beautiful creatures of this former sea. To help identify the invertebrates, Mantell wrote to James Sowerby, a naturalist who was compiling a catalogue of fossil shells. There were many different species of ammonite, the extinct mollusc with a spiral shell that had so enchanted him as a child. In gratitude for the perfect specimens Mantell sent, Sowerby named one species after him: *Ammonites mantelli*. Embedded with the ammonites in the chalk were bivalves, sponges and another extinct mollusc, the belemnites, with their characteristic conical shell divided into chambers.

As Mantell gained in confidence, he established a network of correspondence with those of similar interest, such as the indefatigable Etheldred Benett – a woman of formidable intellect and serious endeavour, not one to swoon with delight at fashion's latest dictates or, indeed, to follow them. Although from a prosperous family in Wiltshire, Miss Benett did not succumb to the usual conventions of a woman in county society. Rather than settling into marriage in some comfortable country rectory she stayed firmly unattached, and her pony carriage would often be seen on the hills of Wiltshire while she pursued her main interest, geology. She devoted her life to her collection, and became so well known that eventually her name alone in the literature would suffice to denote work of outstanding quality.

For Mantell such correspondence was both prestigious – she was, after all, a member of the gentry – and highly beneficial, since he could extend his knowledge beyond Sussex. He wrote to her in 1814, requesting 'the honour of a correspondence'. Miss Benett graciously replied by sending a hamper of fossils to the wagon office in Lewes. Soon they were immersed in comparing the strata of Wiltshire and Sussex, trying to decide which rocks lay above and below others in the sequence. Although not yet known as such they were trying to unravel the sequence of 'Cretaceous' rocks across Southern England, which had been formed between 144 and 66 million years ago. Mantell was aided in this by another member of the gentry, George Greenough, who was busily engaged in developing his geological map of England with William Buckland. Greenough was only too keen to capitalise on Mantell's enthusiasm, frequently requesting more detailed information on the Sussex rocks. By 1815, Mantell had already identified several different strata within the chalk formation around Lewes, the lowest being blue marl, then chalk marl, lower chalk and upper chalk. Greenough steered his research and provided advice on naming the rocks.

In the course of his medical duties, Gideon Mantell was summoned to the assistance of a Mr George Woodhouse, a prosperous gentleman who owned a linen-draper's business in London. While giving 'unremitting professional attention' to Mr Woodhouse, Mantell could hardly fail to notice his patient's eldest daughter Mary. Her portrait shows a young woman with a mass of dark curls piled high, and large, regarding eyes, her face set off by a fashionably off-the-shoulder dress. She shared his interest in fossils and gave him gifts – corals from Worcestershire and other curiosities she had found. They soon 'formed an attachment' and, it would appear, could not wait to get married.

The bride was only twenty, a minor in the eyes of the law, when she married in May 1816, by special licence and with the consent of her

Mary Mantell.

mother as guardian. Unfortunately, her father did not live to see the
wedding. Thanking Mantell for his 'professional exertions and kind
attention' shortly before he died, Mr Woodhouse presented him with
a treasured gift of James Parkinson's survey of the fossil kingdom.

Once settled in Lewes, Mary Mantell dutifully helped her husband with the painstaking task of searching for fossils, which was rapidly becoming his consuming interest. It was not uncommon for her to ride out with him on geological expeditions and sometimes even on his medical rounds, when she would check the ground for fossils while he visited patients. She soon found she could assist him with drawings of his finds, and patiently tried to master the art of scientific illustration. 'I am much pleased with her [Mary's] first attempt at etching,' wrote Etheldred Benett to Mantell in 1817. 'A little practice to enable her to work stronger and bolder appears to me all that is wanting to make them a great ornament to your work.'

With growing self-confidence in his geological observations, Gideon Mantell now decided to write a book setting out his findings on the rocks of Sussex, which he hoped would establish his scientific credentials and perhaps secure his membership of one of the prestigious scientific societies. Mary undertook the illustrations: a fragment of the claw of a crustacean, part of the dorsal fin of a fish, the extraordinary sharp spines of the *Plagiostoma spinosa* – all the dismembered bits of Nature in their incredible variety. With his wife's total support and interest, Mantell described his happiness after his marriage as 'greater than ever'.

As Gideon Mantell began to explore further afield, he realised that there was very different rock to be found in an area known as the Weald, a forest-covered ridge lying between the chalk hills of the North and South Downs. 'Advancing from the Downs, an outcrop of sandstone is first seen near Taylor's bridge,' he observed, 'and it subsequently appears by the stream that winds along Cuckfield park. At this spot in Whiteman's Green an excavation has been made.' As he approached the quarry, he could see tucked below the gorse, wild thyme and trees clinging to the rocks at the surface, the Weald strata exposed to a depth of some forty feet. There were horizontal layers of sandstone, limestone and slate, lying on a bed of blue clays.

With growing excitement, as he examined fragments of rock he began to realise that the fossils entombed in the layers of sandstone and lime-stone at Whiteman's Green were quite unlike the invertebrates of the chalk hills around Lewes. Embedded among the debris appeared to be petrified fragments of *larger* bones. He mentioned this in his letters to George Greenough and Etheldred Benett in the autumn of 1817, explaining that he had uncovered the teeth and bones of vertebrates: amphibia, perhaps crocodiles or alligators. However, the fossils were so mutilated and worn that they were almost impossible to interpret or classify. The logical, tidy plans of the past months, the neat orderly drawings of invertebrates, everything that had fired his imagination about the former sea of chalk, began to be overshadowed by these curious fossil beds.

Whiteman's Green was too far from Lewes for Mantell to ride out each day, and with the arrival of his first child, Ellen Maria, in 1818 he had even less time than before. So he began negotiations with a Mr Leney who ran the quarry, and during the next year parcels from Cuckfield began to arrive at the Lewes wagon office. The first delivery was not particularly exciting: 'the bones, teeth and the tongue of a fish'. After another abortive trip when 'it rained in torrents nearly the whole of our journey', Mantell 'made further arrangements with Leney respecting the Cuckfield fossils'. He is likely to have taught the quarryman to search for the remains of larger bones. It wasn't long before several packages arrived from him, including some fossils that Mantell described as 'superb specimens'.

Among the fragments of larger bones there were also invertebrates, such as tiny fossil shells and snails. Mantell tried to describe these to Etheldred Benett, although he admitted they were so damaged 'it is scarcely possible to ascertain the genera or species'. Nonetheless, Miss Benett thought the shells from the Weald were similar to those uncovered in a rock she knew called 'Purbeck limestone'. This is a formation that stretches across Wiltshire and Dorset that was well established in the geological sequence as Secondary rock. The Weald and the Purbeck, Mantell wrote, following Miss Benett's advice, 'correspond in so many particulars . . . that there is every reason to suppose that they belong to the same formation'. If this was true, the rocks at Whiteman's Green in the Weald lay well below the chalk at the top of the Secondary series. He was revealing tantalising glimpses of a former world that had thrived an unknown number of years before the fishes and ammonites embedded in the chalk.

All this time Mantell's medical practice prospered, and he was able buy a house in Castle Place from his former partner, James Moore, for £700. By 1819 he could afford the house next door and the two houses were knocked together, becoming known, grandly, as 'Castle Place'. Positioned prominently in the High Street and backing on to Lewes Castle, the imposing home was a world apart from the modest cottage in St Mary's Lane where Mantell had been brought up. A team of craftsmen was hired to create a gracious interior with Georgian windows to the floor, ornamental arches over the stairwell and even carvings of ammonites to decorate the Ionic columns at the front of the house. As if to complete the metamorphosis from bootmaker's son to doctor of standing, Mantell adopted the coat-of-arms of his forebears and painted them, entwined with those of Woodhouse, on the porch outside and on the marble table in the dining-room. But if Mary Mantell was under any illusion that she might acquire an elegant new drawing room to entertain guests, she was to be disappointed. Her husband's burgeoning 'little cabinet' became a grand 'Collection' and quickly came to fill the new first-floor drawing-room.

As news of Gideon Mantell's collection spread, visitors came to view the fossils. One caller was that same Lieutenant-Colonel Thomas Birch who had sought out Mary Anning; Mantell described him as a 'very agreeable and intelligent man'. Birch had been touring the West Country and had spent much time in Dorset buying fossils of the *Ichthyosaurus* – or 'Proteosaurus' as it was still called – from Mary. Naturally, Mantell was intrigued to know how Birch's giant *Ichthyosaurus* bones compared to the fragments he had found. In March 1820, shortly after the birth of his second son, Walter Baldock, Mantell received an intriguing letter from Lieutenant-Colonel Birch.

'I am going to sell my collection for the benefit of the poor woman and her son and daughter at Lyme who have in truth found almost *all* the fine things,' Birch wrote. 'I found these people, the Annings, in considerable difficulty – on the act of selling their furniture to pay their rent – in consequence of their not having found one good fossil for near a twelvemonth. I may never again possess what I am about to part with; yet in doing it I shall have the satisfaction of knowing that the money will be well applied.'

Birch was genuinely concerned that the Annings had not been able to maintain their early success. Apart from an *Ichthyosaurus* uncovered in 1818, they had had no more significant finds. Birch's sale was planned for 15 May, in the Egyptian Hall in Piccadilly. Gideon Mantell attended the auction and had a chance to see Mary Anning's 'marine lizards': an *Ichthyosaur* femur and head that was bought for Georges Cuvier in Paris, a partial skeleton, vertebrae and other fossils. Lieutenant-Colonel Birch's sale raised £400 for the Annings.

Less than a month later, in June 1820, Gideon Mantell entered in his diary: 'received a packet of fossils from Cuckfield. Among them was a fine fragment of an enormous bone; several vertebrae and some teeth.' Having met Birch and seen Mary Anning's giant sea lizards, he immediately wrote that these giant bones must belong to a 'Proteosaurus or Ichthyosaurus.' After all, this was the only large creature that had been described in England. Inspired by this discovery of the largest bone he had received so far, he began a series of excursions to Whiteman's Green in the Weald, the mundane little quarry where workmen laboured for basic stone material, unaware that they were laying bare the secrets of the past. To Mantell the quarry was a magical place; it was like entering the ancient tombs, where extraordinary records of a former world were waiting to be explored. With great enthusiasm, on 16 August 1820 he took the entire family on an outing to the quarry: 'We made an excursion to Cuckfield; my brother drove the ladies in his chaise and I rode on horseback.'

But the more specimens he found, the more baffling the site became. Strangely, although he believed the animal bones belonged to an *Ichthyosaurus*, or sea lizard, he began to find the petrified remains of land plants. These fossils were difficult to interpret, some blackened like charcoal, with cracks and fissures filled with white crystalline minerals or the brilliant bronze of fool's gold. As he scrutinised the stone, he thought he could discern fragmentary remains of leaves, stems and other 'ligneous structures', which appeared to be of vegetable origin.

Before 1820, very little was known of fossil botany. In the eighteenth century, Carl Linnaeus had developed a detailed classification system for plants, establishing in his catalogues of several thousand plant species from all over the world the ground rules that a botanist should follow to describe and name plants correctly. Since then, other scholars had occasionally attempted to identify fossil plants, but there were few systematic studies of fossil species before 1820, and names had no legal status. Faced with tantalising impressions of the relics of plants that he could not recognise, Mantell had no knowledgeable source to which he could turn. 'I am unacquainted with any vegetables either recent or fossil with which these remains can be identified,' he wrote.

During one trip to the quarry at Cuckfield in 1820 he had a break-through. He unearthed, buried with more giant bones, part of a tree trunk more than three feet long, very weathered, with the rudiments of branches. He could see at once that the trunk was covered in distinctive diamond-shaped scars, resembling woody bases where leaf stalks were once attached. This was nothing like the English trees around him, the familiar indentations on the bark of oaks, chestnuts and birches. The roughened surface of the trunk, the pattern of scars from woody leaf stalks, were striking – like those of a tropical palm.

Mantell soon found other fossils, too, which bore more resemblance to a tropical flora. Some of the leaves and stems he thought were like *Euphorbia* from the East Indies, a lush, flowering shrub. With some confidence he entered in his journal on 17 August 1820: 'Had a very fine specimen of Euphorbia from Cuckfield.' Two weeks later, he sent 'a large and beautiful specimen of fossil Euphorbia from Cuckfield to Mr Greenough . . . it was embedded in mastic, the same composition as used for the Minerets and Domes of the oriental palace at Brighton'. In fact, flowering plants, like *Euphorbia* and palms, had not yet appeared on the map of the primitive landscape. The history of fossil plants and the habitat for Mantell's unknown giant creature were stranger than anything he could anticipate with the limited evidence then available to him.

In 1821, as Mantell was trying to find out more about tropical plants and animals, the Reverend William Conybeare completed his detailed study of *Ichthyosaurus* for the Geological Society. This was to provide another clue to the giant bones. Conybeare had included beautiful anatomical drawings of the bones of *Ichthyosaurus*. When Mantell compared the fossil bones that he had uncovered at Whiteman's Green in the Weald, he found that they were very different from those of the sea lizard of Lyme. The vertebrae of an ichthyosaur were slender and deeply hollowed, allowing for the flexible movements of an animal living in water – nothing like the chunky, solid vertebrae that he had uncovered in Sussex. The leg bone of the *Ichthyosaurus* was more like the fin of a fish; the slender central bone, the humerus, 'immediately supporting a very numerous series of small bones, form[ed] a very flexible paddle'. But the portion of giant femur, or thigh bone, that he had found in the Weald bore no resemblance to any bone in the sea lizard. It was truly enormous: the fragment, of the top part of the bone, was over two feet long and twenty inches in circumference. If this fragment of a giant leg was not derived from an *Ichthyosaurus*, then to what kind of monstrous creature *could* it belong?

Apart from the shape of the bones, there was another clue that the unknown creature from the Weald was not a sea lizard. When a creature dies at sea, its body sinks down to the ocean floor and is gradually covered by a fine rain of particles that form the new sediments. As it is gradually densely packed with layers of sediment accumulating above, the bony skeleton can be well preserved, just like the ichthyosaurs of Lyme. But when a creature dies on land it is much more likely to be destroyed, falling prey to some other animal or scattered by wind and rain, leaving only a confused jumble of bones. Mantell could recover only fragments of bone from the Weald, never a full skeleton. As yet, he had not even found two bones joined together. It occurred to him that these worn relics of giant bones might have belonged to a creature that spent at least part of its life on land, beneath the shade of palms.

In the quiet of night, when all the town was long since asleep and his medical duties were completed, Gideon Mantell studied the fossils he had found, so utterly absorbed in his work that he was often unaware that the small hours were approaching. With careful use of chisel and hammer, the shape of the bones slowly emerged from the surrounding stone like some strange primordial sculpture, perhaps more impressive than something that is finished, containing all the promise of a great work of art gradually taking shape before his eyes. He would glimpse eerie fragments of the ancient animal: the exquisitely smooth curve of the giant femur, the sharp points of the damaged vertebrae, the strange ridges on the enamel of the teeth; the foramina, or holes, for blood vessels, far larger than any human capillary. It was unearthly.

To try to make more sense of this confusing picture, he would use as a reference Georges Cuvier's acclaimed four-volume summary, *Recherches sur les Ossemens Fossiles des Quadrupèdes*, which had been translated into English in 1813. Here Cuvier outlined the details of several species of ancient extinct crocodiles found at Honfleur and Le Havre. Mantell compared his fossils against Cuvier's drawings, and some of the bones, especially the vertebrae, seemed to correspond. To obtain a second opinion, he now made arrangements to view the Hunterian Museum at the Royal College of Surgeons in London. John Hunter's collection of ten thousand anatomical specimens had been bought by the government after his death in 1793 and placed in the Royal College at Lincoln's Inn Fields. Here, they were being catalogued by Hunter's former apprentice, William Clift.

The child of a poor family in Devon, Clift possessed an exceptional talent for drawing which had been noticed by the local gentry, and he had been sent to assist Hunter. Greatly honoured by this appointment, the young Clift had laboured long hours for little pay, helping with dissections before breakfast at six in the morning and often not finishing until after midnight, the evenings being spent in dictation. Within a year of his apprenticeship, Hunter died, but he revered his former master and was determined to continue his work.

Although hampered by the fact that Sir Everard Home had removed many manuscripts that would have helped to identify the specimens, Clift struggled on, trying to prepare Hunter's collection for public view.

By the 1820s, his experience was considerable. When Gideon Mantell presented him with one of the pointed, curved teeth he had found, Clift did not hesitate: 'there can be no doubt of its having belonged either to the crocodile or the monitor [lizard]. I know of no animal whose teeth have the lateral ridges so strongly defined.'

From such discussions with Clift and comparisons with Cuvier, Mantell began to think that at least some of the giant bones could be assigned, not to a sea lizard, but to an ancient species of crocodile. He wrote in summer 1821 to a friend, the MP and Fellow of the Royal Society Davies Gilbert, telling him of the giant bones of crocodiles that he had found that spring in the Weald. 'There can be no hesitation,' said Mantell confidently, in assigning them to 'the same unknown species of Crocodile, as discovered at Honfleur and Havre'.

But soon after this, Mary Mantell made a remarkable discovery that did not fit this conclusion. There are several versions of the event; the most plausible recounts that the incident occurred one morning in 1820 or 1821, when Mary was accompanying her husband on his medical rounds. While waiting for him to see his patient, she searched for fossils. As she walked, her eyes were irresistibly drawn to a strange shape in a pile of stones that had been heaped by the side of the road. Picking up the stone, she brushed away the white dust, gently removing any loose rock with her fingers. Gradually a shape emerged never previously seen by human eye. It was very smooth, worn and dark brown, rather like a flattened fragment of a giant tooth.

When she showed her husband, he saw at once that this was something important. 'Soon after my first discovery of colossal bones,' he wrote, 'some teeth of a very remarkable character particularly excited my curiosity for they were wholly unlike any that had previously come under my observation.' The fragment of tooth was more than an inch long, and shaped into a blunt, grinding surface at the crown. The couple were able to trace the source of the pile of stones to the same quarry in Whiteman's Green in which Mantell had found the other giant bones. 'Even the quarrymen, accustomed to collect the remains of fishes, shells and other objects embedded in the rocks,' he wrote, 'had not observed fossils of this kind and were not aware of the presence of such teeth in the stone they were constantly breaking for the roads.'

The 'tooth' cast all his observations into doubt. He could see that this was not the tooth of a crocodile, for it did not have the sharp, pointed crown essential for a carnivore. It had a broad, flattened grinding surface, supported by thick enamel on one side and with a marked ridge up the middle. This was much more like the tooth of a herbivorous mammal, that had been worn down by constant chewing. 'The first specimen so entirely resembled the part of the incisor of a large mammal,' he wrote, 'that I was much embarrassed to account for its presence in such ancient strata, in which according to all geological experience, no fossil remains of mammal would ever be discovered.'

Although the tooth resembled a herbivorous mammal like a hippopotamus or rhinoceros, such creatures were not supposed to exist in ancient rock. Cuvier's extinct large mammals such as the mammoth and mastodon had been retrieved from Tertiary deposits. Mantell thought from his correspondence with Etheldred Benett that Weald rocks were much older, from the Secondary period. To suggest that mammals had lived in ancient times was one step beyond anything that naturalists could envisage. As James Parkinson had written, although the time-scale of Creation as outlined in Moses had been questioned, the order of Creation was not in doubt. Parkinson thought it striking that the order of creation as stated in the Scriptures was 'in close agreement' with geological evidence: 'The Creative Power has been exercised with increasing excellence in its objects . . . the last and highest work appearing to be Man.' No one had yet challenged the assumption that mammals were created last, when God had prepared the Earth for the higher animals.

This belief informed Mantell's quest; he did not yet have enough evidence to disregard the huge burden of accepted wisdom. He asked himself, if the owner of the tooth was not a large mammal, then what was it? The tooth did not resemble that of any fish at the Hunterian Museum. It could not come from a turtle; they have no teeth, only horny beaks. No amphibian was known to reach giant proportions. And it certainly was not from a bird – no toothed birds had been reported at this time. By a process of elimination the evidence pointed to a bizarre conclusion: the teeth belonged to a giant herbivorous lizard.

Yet this conclusion made no sense. 'As no known, existing reptiles are capable of masticating their food I could not venture to assign the tooth in question to a lizard,' wrote Mantell. A herbivorous reptile that could chew its food like a cow was unheard of. It was a preposterous idea. The experts in London, such as William Clift, were following Georges Cuvier, interpreting the fossil record by analogy to living forms. But there was no modern analogue to such a strange reptile.

Gideon Mantell lacked the one piece of evidence that would have proved the tooth belonged to a reptile: a fossilised jaw. A mammal's jaw is very distinctive. Even if the teeth are missing, there are differently shaped spaces for the various types of teeth: molars, premolars, incisors and canines. A reptile does not have several types of teeth; although its teeth may vary in size, the sockets are all the same shape. But Mantell could not find a jaw, just a single disembodied tooth.

When he studied the tooth at home, in his drawing-room, surrounded by his collection, in moments of doubt – the large fragment of tooth was so worn – he sometimes wondered if he had found anything at all. Viewed from some directions it was almost unrecognisable as a tooth. Fine, feathery black lines were woven across the surface like a spider's web. There it lay in his hand, a scrap of a fossil scarcely larger than a pebble, withholding the secret to an unknown past.

During the summer of 1821 Mantell redoubled his efforts to gather any evidence that could shed more light on the mystery. Scarcely interrupted by the major events of the day – the death of Emperor Napoleon on St Helena, the spectacular coronation of the ageing King George IV, the summer races and Brighton fair – he struggled with his geological research whenever time could be spared from his practice. Sometimes he took the single-horse chaise to Cuckfield with his young apprentice, George Rollo; occasionally he rode out alone to hunt for further evidence of his monster.

By the autumn, Mantell's first-floor rooms were filled with a strange assortment of fragments of giant bones uncovered in the Weald. From his knowledge of anatomy while training as a doctor he was able to identify several of them. He wrote to the Reverend Conybeare at the Geological Society, telling him that he had 'ribs; clavicle [part of the shoulder]; radius [forearm]; pubis [front part of the pelvis]; ilium [from the side of the pelvis]; femur or thigh bone; tibia or shin bone of the leg; metatarsal bones of the foot; vertebrae forming the back-bone and teeth'. Although the teeth appeared broken off close to the jaw, the jaw itself could not be traced. Some of the bones had features in common and seemed to belong together. Others were so broken and fragile they were impossible to identify. All the bones were hopelessly intermingled with debris from other animals, turtles, fishes, shells and vegetables.

The only way he could begin to make sense of the puzzle was to distinguish the different types of teeth. There seemed to be two sets here that could not have come from the same species of animal. One set of teeth were blade-like and up to three inches long, flattened from side to side, with two sharp edges stretching from the crown. These edges were serrated like a steak knife, constructed for tearing flesh, not eating vegetables. The teeth could only have belonged to a carnivorous animal. And although he couldn't prove it beyond doubt, Mantell was certain that they belonged to a giant reptile because they were more similar to crocodile teeth than anything he had seen at the Royal College of Surgeons. However, there were some crucial differences. Crocodile teeth are conical, slightly curved, the surface of the enamel covered with ridges radiating longitudinally from the tip to the crown. A crocodile grips its prey, and then flicks its tail in the water to spin, so it can more easily rip and tear off chunks. These unfamiliar, blade-like, carnivorous teeth would have allowed their unknown owner a slicing action, like carving meat.

Even more puzzling was the second set of teeth in his collection, the herbivorous teeth found by his wife. These 'possessed characters so remarkable that the most superficial observer would have been struck with their appearance as something novel and interesting,' he wrote. 'When perfect they must have been of a very considerable size.'

Self-taught, without the backing of a university or membership of a prestigious society, Mantell could hardly claim that these once belonged to a giant herbivorous lizard when such an improbable creature was not supposed to have existed. He might just as well suggest he had found a centaur, a unicorn or a dragon, or some other preposterous creature of ancient myth.

But the most remarkable feature of all was the sheer size of the beasts. Some of the fragments of vertebrae were up to five inches long; there was a part of a rib that measured twenty-one inches long, even the metatarsal bones in the foot were huge and chunky. As he was chiselling away one night, he realised that one particular broken section of thigh bone emerging from the stone indicated an animal far larger than any he knew – this piece was nearly 30 inches long and 25 inches in circumference. There it lay in front of him, defying all logic and reason. There was no way of proving to which set of teeth it belonged. Compared to a mammal bone, if scaled up in size his discovery would make a preposterous animal, far larger even than a house.

> I may be accused of indulging in the marvellous, if I venture to state that upon comparing the larger bones of the Sussex lizard with those of the elephant, there seems reason to suppose that the former must have more than equalled the latter in bulk and have exceeded thirty feet in length! And yet some bones in my possession warrant such a conclusion . . . this species exceeded in magnitude every animal of the lizard tribe hitherto discovered, either in a recent or a fossilised state.

Could a heart really pump blood around a creature 30–40 feet long? Would muscles be strong enough to support such a heavy frame? What would it have eaten to keep its several tons of reptile flesh in pristine vigour? The creature beginning to emerge from his solitary work each night was hardly believable, a phantom from the underworld, yet there it was, solid as a rock, unassailable. As a glimpse of an ancient form of life it was tantalising; a seemingly endless, uncompletable, jigsaw. None of it added up to a whole animal, or even a coherent view of part of an animal. But with single-minded, purposeful dedication, Mantell continued to devote all his spare time to trying to solve the mystery. Everything in his life was sacrificed to this one bewitching interest. *He* would place the bones in history. *He* would be the man acclaimed.

But unknown to Gideon Mantell, he was not the only person in England in the early 1820s who had uncovered evidence that giant lizards once roamed the land.

3

Toast of Mice and Crocodiles for Tea

Here we see the wrecks of beasts and fishes
With broken saucers, cups and dishes . . .
Skins wanting bones, bones wanting skins
And various blocks to break your shins.
No place in this for cutting capers,
Midst jumbled stones and books and papers,
Stuffed birds, portfolios, packing cases
And founders fallen upon their faces . . .
The sage amidst the chaos stands,
Contemplative with laden hands,
This, grasping tight his bread and butter
And that a flint, whilst he doth utter
Strange sentences that seem to say
'I see it all as clear as day.'

'A Picture of the Comforts of Professor
Buckland's rooms in Christ Church,
Oxford' by Philip Duncan, 1821, cited in
*The Life and Correspondence of William
Buckland* by Anna Gordon, 1894

In the heart of Oxford, under the watchful eye of the deans and canons at the university, the Reverend William Buckland's enthusiasm for 'undergroundology' was beginning to attract wider support. As Reader in Mineralogy he had expanded the course to debate the latest geological ideas: whether the 'days' of Creation could correspond to lengthy 'eras'; the nature of Noah's Flood; the order of Creation. According to one reviewer, Buckland was so inspiring as a speaker that 'he awakened in the University and elsewhere, an admiration and interest in Geology'. He told his friend the amateur geologist Lady Mary Cole that he had been lecturing to an 'overflowing class . . . amongst whom I reckon the Bishop of Oxford, four other Heads of Colleges and three Canons of Christchurch'.

His idiosyncrasies were becoming almost as famous as his lectures and were accepted at the university as part of his brilliance. Anyone passing through the neatly trimmed rose gardens of the quad at Corpus Christi to Buckland's rooms, expecting to find the usual happy amalgamation of elegance and learning fitting for a don, would soon discover that the professor had different priorities. 'I can never forget the scene that awaited me on repairing from the Star Inn to Buckland's domicile,' recalled Roderick Murchison, an undergraduate at Oxford. 'Having

William Buckland lecturing at the Ashmolean, 1822.

Awful Changes.
Man found only in a fossil state. —— Reappearance of Ichthyosauri.
"A change came o'er the spirit of my dream." Byron.

A Lecture. — "You will at once perceive," continued Professor Ichthyosaurus, "that the skull before us belonged to some of the lower order of animals the teeth are very insignificant the power of the jaws trifling, and altogether it seems wonderful how the creature could have procured food."

A cartoon of 'Professor Ichthyosaurus' inspired by Buckland's lectures.

climbed up a narrow staircase . . . I entered a long corridor-like room filled with rocks, shells and bones in dire confusion. In a sort of sanctum at the end was my friend in his black gown, looking like a necromancer, sitting on a rickety chair covered with some fossils, clearing out a fossil bone from the matrix.'

In addition to fossils strewn liberally on almost every surface and the stuffed creatures in the hall, Professor Buckland was a keen naturalist and kept a number of unusual pets. There were cages full of snakes and green frogs in the dining-room, where the candles were placed in *Ichthyosauri*'s vertebrae. Guinea-pigs roamed freely throughout his office. Walter Stanhope, a tutor at Oxford, described an evening in

Buckland's apartments: 'I took care to tuck up my legs on the sofa, for fear of a casual bite from a jackal that was wandering around the room. After a while I heard the animal munching up something under the sofa and was relieved that he should have found something to occupy him. I told Buckland. "My poor guinea pigs!" he exclaimed, and sure enough, four of the five of them had perished.'

By far the most splendid creature in Buckland's menagerie was a bear, rather grandly named Tiglath Pileser, after the founder of the Assyrian Empire in the Old Testament Book of Kings. Unlike his name-sake, who was renowned for his brutal punishment of his opponents, Tiglath the bear was 'tame and caressing'. Buckland even went so far as to provide the bear with a student costume in which he participated fully in university life, especially the wine parties. 'We had an immense party at the Botanic Gardens,' Charles Lyell, one of Buckland's under-graduates, recalled. 'Young Buckland had a bear, "Tig" dressed up as a student complete with cap and gown.' Tiglath Pileser was formally introduced to senior figures at the university. 'It was diverting to see two or three of the dons not knowing what to do for fear their dignity was compromised.'

Most perplexing of all for visitors to Buckland's apartments was the menu, since Buckland, a born experimentalist, had decided to eat his way through the animal kingdom as well as study it. 'I recollect various queer dishes which he had at his table,' recalled his friend John Playfair. 'The hedgehog was a good experiment and both Liebig and I thought it good and tender. On another occasion I recollect a dish of crocodile, which was an utter failure . . . though the philosophers took one mouthful, they could not be persuaded to swallow it and rejected the morsel with strong language.' John Ruskin, recalling his undergraduate days at Buckland's table, wrote: 'I met the leading scientific men of the day, from Herschel downwards . . . Everyone was at ease and amused at that breakfast table, the menu and the science of it, usually in themselves interesting. I have always regretted a day of unlucky engagement on which I missed a delicate toast of mice.'

The discussions that graced these gastronomic occasions were undoubtedly no less exotic. Buckland believed that geological history reflected a gradual preparation of the earth for Man's habitation and was optimistic that a scientific history of the earth would tally with scriptural records. He was impressive in debate and was soon influencing some of the more liberal churchmen of his day. John Bird Sumner, the Bishop of Chester and later Archbishop of Canterbury, wrote a *Treatise on the Records of Creation* in 1816, in which he supported Buckland and other members of the Geological Society in viewing the six 'days' as six creative 'eras'.

Buckland's keenness to reconcile the new science with religion won him support in high places. As his reputation grew, he made the acquaintance of leading gentlemen of the day, including Lord Grenville, the Chancellor of Oxford University; Sir Joseph Banks, the famous botanist; and Sir Everard Home at the Royal Society, as well as leading politicians such as Robert Peel. Using these powerful contacts, Buckland lobbied for the first chair of geology to be created at Oxford. He reassured Lord Grenville that the sciences would, of course, be subordinate to the classics. 'I would not surrender a single particle of our system of classical study,' he promised. The matter was referred to the highest level of government, eventually reaching His Royal Highness, the Prince Regent.

In 1818, with the approval of His Royal Highness, the stipend for a Professor of Geology at Oxford was allotted from the Treasury. 'I feel quite proud of the high consideration which is given to the noble subterranean science by such exalted personages,' Buckland told Lady Mary Cole at Penrice Castle. However, such approval from leading members of society added to the pressure on Buckland to satisfy the urgent need to find geological evidence that would corroborate the Scriptures, such as a biblical Flood. The religious tradition was so entrenched at Oxford that if geologists could not discover such evidence quickly, the infant science would lack credibility.

When Buckland became Reader in Geology he also became Director of the Ashmolean Museum. Directly under his supervision in this museum, on display in the heart of Oxford for well over a century, were the bones of an unknown giant animal. As early as 1677 the first Keeper of the Ashmolean Museum, a Dr Robert Plot, had described them. While writing a *Natural History of Oxfordshire*, Dr Plot had come across an inexplicably large portion of thigh bone from a local quarry, weighing more than twenty pounds. He had suspected it was the bone of an elephant brought to England during the Roman invasion of Britain. When later he had an opportunity to study the skeleton of an elephant, he was puzzled to find that the huge Oxford fossil was totally different. There seemed only one conclusion to be drawn. He wrote, the fossil 'has exactly the figure of the lower most part of the Thigh-bone of a Man'.

During the eighteenth century, more giant bones had been discovered in quarries around Oxford. Joshua Platt, a 'Curiosity-Monger', found three large vertebrae buried at Stonesfield, near Woodstock. Later, the same dealer reported part of a giant thigh bone almost thirty inches long which he valued at four shillings, and a fragment of scapula, or shoulder bone. Early in the next century Professor Kidd, Buckland's predecessor as Reader of Mineralogy, had studied the bones and concluded they were derived from some strange mammal. William Buckland did not record any conclusions about the unknown creature in 1818 when he became the Keeper of the museum, although it is likely that people looked to him for an opinion. Impossible to classify and the subject of the wildest speculation, the bones were at once familiar and accepted as everyday objects and at the same time represented a past of incomprehensible strangeness.

However, later that year there was an opportunity for Buckland to extend his unique brand of English hospitality to a very distinguished French visitor: Georges Cuvier. Cuvier was updating his extensive survey of fossils, *Recherches sur les Ossemens Fossiles*, and hoped to see the latest discoveries of giant bones in Oxford. By now, he had almost legendary status throughout Europe. Approaching his fifties, his thick red hair long since dulled, the 'Napoleon of Intelligence' made a powerful impression and the self-confidence amassed from a lifetime of invariably being 'right' was palpable. It was said of Cuvier that his library — containing some nineteen thousand volumes — was so familiar to him that he could remember everything and retrieve any volume or monograph he required in seconds. He had been showered with awards, named Councillor of State in 1813, and was later granted the honorary title of Baron.

Cuvier visited the Ashmolean and was presented with a variety of giant bones: teeth, vertebrae, ribs, part of an enormous thigh bone and confusing fragments of other bones. No two bones, except for some of the vertebrae, had been found connected together. It was impossible to tell from the detached bones whether they originated from different animals of various ages and sizes or belonged to the same creature. Although there are no records of the conversation that took place between Cuvier and Buckland in 1818, subsequent letters between the two reveal that in no time Cuvier had solved the puzzle.

The first clue available to him came from the rocks themselves. The bones from Stonesfield were found in rock at a considerable depth below the surface. The stone was being mined to provide roofs for new buildings, and could only be obtained by going deep underground. 'They descend by vertical shafts through a solid rock . . . more than 40 feet thick, to the slaty stratum containing these remains,' wrote William Buckland. The giant bones 'are not lodged in fissures and cavities but are absolutely imbedded in a deeply situated stratum . . . which extends across England from near Stamford in Lincolnshire to Hinton near Bath'.

Buckland had studied these rocks and confirmed the earlier work of the surveyor William Smith that the Stonesfield slate lay immediately above a stratum known in the geological sequence as 'the oolitic limestone' of Bath. The oolitic limestone was correctly seen as ancient, formed at the same time as the 'Jura [Jurassic] limestone' strata found on the Continent, well below the chalk in the Secondary series. No mammals had been found this far back in the geological sequence; Cuvier's large mammals were found in the more recent, Tertiary formations. So although the thigh bone had mammalian characteristics, with a thickset, straight vertical shaft, Cuvier examined the bones confident that they were far more likely to be from a reptile than a mammal.

Unlike Gideon Mantell's discoveries in Sussex, the huge teeth displayed at the Ashmolean were still attached to the jaw, and this too provided several important clues. Although the holes for the teeth varied in size along the length of the jaw, they were all the same shape, typical of a reptile. Tiny pointed teeth were poking through the jaw beside the adult teeth which, since reptiles have replacement teeth growing through the jaw all their lives, also indicated that the jaw belonged to a reptile. 'The exuberant provision in this creature,' Buckland wrote, 'for a rapid succession of young teeth to supply the place of those which might be shed or broken is very remarkable.' Convinced the bones belonged to a reptile, both from the age of the rocks and the characteristics of the jaw, Cuvier could pronounce with some certainty that it had other reptilian characteristics: it had been oviparous, or egg-laying and had a dry, scaly skin.

But it was much harder to define what *kind* of reptile or lizard it might have been. Cuvier could see that, within the reptile class, it was not like a turtle, because there was no shell and it lacked the distinctive shape of skull and form of vertebrae. The largest reptile known at this time was a crocodile. These bones shared some features in common with crocodiles: the double-headed ribs, the vertebrae with flat articulating surfaces; and the giant thigh bone had a fourth trochanter, an extra surface for muscle attachment. Mammals have only three surfaces for muscle attachment at the top of the thigh bone; crocodiles, like the unknown creature, have four, denoting a tremendous muscle structure. However, there the similarity ended.

Unlike the conical ridged teeth of the crocodile, these teeth were compressed, with a long serrated edge along the whole extent of the enamel, like a steak knife. The exterior surface of the jaw had distinct cavities for the passage of blood vessels and nerves, allowing the creature a very good blood supply to support the activity of the jaw. And whereas a crocodile jaw is long, thin and pointed, this fragment of lower jaw was short, high and narrow, flattened from side to side. From the absence of curvature on any piece of the lower jawbone, nearly a foot in length, it seemed likely that this creature's jaw terminated in a flat, straight, and very narrow snout. Cuvier concluded that of all living animals, these

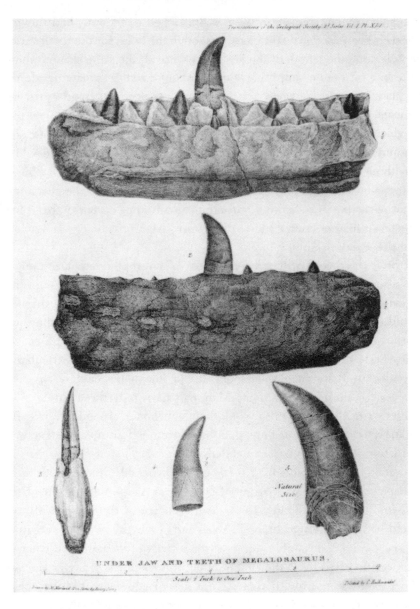

UNDER JAW AND TEETH OF MEGALOSAURUS.

Scale ¼ Inch to One Inch

The jaw of the carnivorous beast in the Ashmolean showing
young replacement teeth – a characteristic of a reptile.

bones were most similar to a carnivorous lizard known as the monitor lizard. However, there was one crucial difference: size. Comparing the thigh bone, which was ten inches in circumference, to the equivalent bone in a lizard, he simply scaled up. 'From these dimensions,' wrote Buckland, 'a length exceeding 40 feet and a bulk equal to that of an elephant seven feet high, have been assigned by Cuvier to the individual to which this bone belonged . . . we may with certainty ascribe to it a magnitude very far exceeding that of any living lizard.'

Although the archives suggest that Buckland had accumulated all this information from his meeting in 1818 with Georges Cuvier and subsequent correspondence, he was in no hurry to publish the findings. His reluctance to announce the find may simply have reflected a reasonable scientific caution. Unlike the ichthyosaurs that Mary Anning had found at Lyme, the Stonesfield animal was far from complete. But Buckland was also well aware that the Anglican authorities who had helped him obtain his stipend as professor from the Treasury were hoping that he would reconcile any geological discoveries with the Bible. A forty-foot reptile was hardly the ideal candidate. After all, there was no record of such a fantastic, almost mythical creature in Moses' account of Creation.

Rather than devoting his time to combing the quarries for further evidence of his huge reptile, Buckland set his sights on another quest altogether: to discover proof of the biblical Flood. In 1819 he presented his inaugural address in geology at Oxford, '*Vindiciae Geologicae*, or The Connexion between Geology and Religion explained'. With great deference to the classical tradition, he explained why 'no evil should be anticipated' if geology was permitted to serve as 'the handmaid of Religion'. He reassured the bishops and deans in the audience that there would be no opposition between the 'Works' and the 'Word' of God. There was no mention of the giant beast of Stonesfield; instead, Buckland expressed his conviction that the new science was bound to provide evidence of the recent origin of Man and the Great Flood.

By 1819, Buckland thought he had convincing evidence for the Deluge. Accompanied by his friend the geological enthusiast Count Breunner of Vienna, he studied the distribution of quartz pebbles and gravels across England. They traced these gravels 'over the plains of Warwickshire, the Midlands, on some hills in Oxfordshire and in the valley of the Thames . . . to below London'. Later that autumn, Buckland wrote a paper for the Geological Society on 'the evidences of the Recent Deluge', in which he proposed that the fearsome torrents of 'the first rush of the advancing deluge' had swept these gravels across southern England. They had, he thought, retraced the actual path of the Flood.

The nearest source to which the Reverend Buckland and the Count could trace the pebbles was Lickey Hill in Worcestershire: 'they present the same glassy brilliancy of fracture . . . the same small crystals of decomposing felspar throughout'. Consequently they believed the pebbles had originated from Worcestershire and had been 'torn up by the waters of the last Deluge'. As the Flood subsided, 'the weight and force of the immense volume of water . . . excavated the series of sweeping combs and valleys', seen for example from Bath to Stow-on-the-Wold. Although Buckland could find no geological evidence to explain what prompted the Deluge and could not define the dimensions of the tidal wave, he was in no doubt that a giant surge or tidal wave had once occurred.

In pursuing evidence for a Flood, Buckland was hoping to resolve philosophical issues that lay at the heart of geology. This would not only add credibility to the new science but could also shed light on what happened to the 'former worlds' uncovered by geologists. There was, as yet, no framework within which creatures such as the *Ichthyosaurus* or the strange reptile from Stonesfield could be understood. Where did these beasts come from and, above all, what had happened to them? Why had God erased these creatures from the face of the earth? In England, where the Anglican faith dominated academic centres like Oxford, the best clue to extinction was the biblical Flood. But in France, naturalists were beginning to put forward new ideas.

Since the discovery that mammalian species such as the mammoth and the mastodon had disappeared from the earth's surface, the puzzle of extinction had been keenly debated in Paris at the Muséum National d'Histoire

Naturelle. Georges Cuvier and a senior colleague at the museum, the 'Professor of Insects and Worms' Jean-Baptiste Lamarck, had developed radically opposing theories. According to Lamarck, species were not necessarily extinct at all. They had developed by 'transmutation' into other forms of life.

Lamarck's thinking stemmed from eighteenth-century beliefs that all living things were linked by imperceptible transitions; Nature was a continuous 'Chain of Being'. The simplest organisms on the scale were those that maintained the minimum conditions for life, and Man, the supreme form, was at the top of the hierarchy. The great Chain of Being was an attempt to explain the incredible diversity of living forms in the absence of any chronology showing the order in which animals appeared on the earth. Lamarck believed that as organisms in this 'scale of being' strove for perfection they could transform themselves while adapting to their environment. Changing circumstances led to new responses from animals, which eventually became habitual. Organs could change permanently by frequent use or habits, allowing for the progression of animal forms into ever more complex types, *without* any special creation from God. This is what he meant by the 'transmutation' of species. In his *Philosophie Zoologique* published in 1809 he outlined a thesis in which humble creatures could 'generate' into higher forms of life.

Lamarck had little evidence to back up his ideas; the fossil record at the beginning of the nineteenth century was so incomplete that there was no proof of the progression of life over time. From his studies on fossil invertebrates, he could only show that the fossil molluscs such as ammonites and belemnites found in ancient Secondary rock were very different from living species. Neither did he propose a convincing mechanism to demonstrate how evolution might have occurred. Nonetheless, in his lectures he described the invertebrates as the most primitive forms of life and, he speculated, 'perhaps the ones with which Nature began, while it formed all the others with the help of much time and of favourable circumstances'. His ideas on development implied that no species became extinct – they were merely transformed: 'one may not assume,' he wrote in 1802, 'that any species has really been lost or rendered extinct'.

Jean-Baptiste Lamarck's revolutionary thinking had worrying implications. Could intelligence and rational thought, the 'God-given' attributes which set Man apart from animals, have developed from more primitive forms of life? If organisms transformed themselves and higher forms could emerge from lower forms, then Man was not specially made by God. Buckland's friend Conybeare was one of many to denounce Lamarck's 'ridiculous' theory. It was 'an idea so *monstrous'*, Conybeare told the Geological Society in 1821, 'that nothing less than the credulity of a material philosophy could have been brought for a single moment to entertain it, nothing less than its bigotry to defend it'. The idea that Nature was autonomous and could randomly generate higher forms of existence, including Man, was greeted with intense hostility and roundly condemned.

In France, Lamarck had difficulty even in obtaining publishers for his ideas. Cuvier was so antagonistic to this 'evolutionary' thinking, it is thought that he advised the Emperor Napoleon not to accept a copy of Lamarck's *Philosophie Zoologique*. It was a well orchestrated public humiliation. In his lectures, Cuvier scoffed at the notion that organs could be formed by frequent use. He challenged Lamarck's view that the entire animal kingdom was united in one genealogical tree. Cuvier believed that the differences between, for example, a humble mollusc and a complex vertebrate were so great that they could not possibly have arisen from a continuous chain.

Cuvier had developed a different theory to account for extinction, called the 'Doctrine of Catastrophes', according to which violent 'revolutions' had wiped away former worlds, destroying ancient forms of life. These ideas stemmed from a study he had undertaken with another colleague at the Muséum National d'Histoire Naturelle, the Professor of Mineralogy Alexandre Brongniart. Together they made a special study of the conditions under which fossils had become entombed in the Tertiary rocks of the Paris basin. For four years, almost every week, they took the carriage into the countryside around the River Seine.

Georges Cuvier.

Above the chalk of the Secondary strata they identified several major Tertiary formations. Each layer of rock had its own characteristic fossils, some containing marine invertebrates, others only freshwater creatures. These alternating layers of marine and freshwater formations led the two scientists to conclude that there had been repeated incursions of sea. Because there were 'abrupt junctions' between the marine and freshwater formations, they reasoned, the ocean had invaded suddenly, submerging the land for prolonged periods and destroying living species.

The ancient globe, Cuvier reasoned in his *Essay on the Theory of the Earth*, was punctuated by a series of 'revolutions that were so stupendous that . . . the thread of Nature's operations was broken by them and her progress altered'. He envisaged that prior to the creation of Man there were several different periods in the earth's history, shown by the many different layers of rock in the earth's crust that were filled with fossils. Each period ended in a dramatic geological 'catastrophe' in which species became extinct. 'Life has often been disturbed on this earth by terrible events,' wrote Cuvier. 'Numberless living beings have been the victims of these catastrophes; their races have even become extinct.'

When Cuvier's *Essay* was translated into English, the editor, Professor Robert Jameson of Edinburgh University, presented Cuvier's theory as though the most recent 'catastrophe' *was* the biblical Flood. This was an obvious mistranslation of the Frenchman's original ideas, which were based upon research within the Paris basin. Nevertheless, in England this was embraced as authoritative scientific backing for the Bible. William Buckland praised Cuvier's 'inestimable Essay', and was eager to extend his notion of incursions of sea to 'a recent Deluge acting universally over the surface of the whole globe'. He also hoped to show how this might correspond with the layers of rock that formed the earth's crust.

By 1821, Buckland and his friends at the Geological Society had made considerable progress mapping the succession of strata in England. Following William Smith's earlier studies, they identified several major formations in the Secondary series, complementing Cuvier's studies of the Tertiary rock above. There was still little known about the oldest Primary and Transition layers. Nonetheless, Buckland and his colleagues had glimpsed as far back in time as the period now known as 'Devonian', the lowermost Secondary rocks. They called these ancient rocks the 'Old Red Sandstone'. Above this, Buckland identified later rock formations: 'Carboniferous Limestone', succeeded by the 'Coal Measures', 'New Red Sandstone' (Triassic), 'Jura limestone' (Jurassic), and finally the most recent chalk and greensand (Cretaceous). These formations

Popular conception of *The Deluge*, painted by John Martin, 1828.

together made up the major periods of the Secondary series. Sadly for William Smith, when the gentlemen geologists of the Geological Society of London published their map, sales of his own map were cut to nothing. Smith became so poor that at one stage he was even reduced to spending time in a debtors' prison.

Although little was known about the fossils in the different layers, this classification of the Secondary rocks proved to be remarkably accurate and still stands up to scrutiny today. Since, as James Hutton had argued, each layer of rock was formed imperceptibly, the result of gradual erosion and deposition over countless years, this classification lent powerful support to the idea of vast geological epochs before the creation of Man. Buckland was beginning to glimpse distinct periods in which centuries of prehistory buried in the earth's crust could be defined.

Buckland was keen to integrate all these threads of evidence: the succession of strata, Cuvier's 'catastrophes' and biblical records of a

Flood. His opportunity came later in 1821, when quarrymen stumbled upon a cave at Kirkdale in Yorkshire containing ancient fossil bones. He hurried to the site, suspecting this would provide further insights. Surely the animals in the cave had been swept in by the terrifying, swirling Flood waters? What he found was stranger than anything he could have imagined.

Deep into the cave he went, on his hands and knees, the circle of light from a candle allowing him brief glimpses of what lay ahead, the voices of his companions echoing in the ancient silence. Undisturbed for centuries, the cave divided into passages that stretched back two hundred feet into the hillside. At first, all he could see was mud and silt. Gradually, it became clear that the scene was much more gruesome. Partially obscured by stalagmites and stalactites, 'the bottom of the cave was strewed all over, from one end to the other, with hundreds of teeth and bones'. 'Scarcely a single bone has escaped fracture,' he said.

Drawings of the fossils were sent to Georges Cuvier, who confirmed Buckland's suspicions that the bones were from many different animals jumbled together in disarray. These were creatures that never live together: tigers and deer, bears and horses, in addition to extinct species of elephant, rhinoceros, hippopotamus and hyenas. Furthermore, it was hard to envisage how large animals such as elephants could have passed through the two-foot entrance to the cave. Even more puzzling, Buckland observed from the splintered fragments and gnaw marks, all the bones appeared to have been *half-eaten*.

Buckland began to suspect that this was an ancient hyena den; the larger animals had been dragged into the cavern, a portion of the carcass at a time. He imported a hyena from the Cape and compared the gnaw marks on bones eaten by it with those from the caves. He soon wrote jubilantly to a friend, the Reverend Vernon Harcourt: 'Billy [the hyena] has performed admirably on shins of beef, leaving precisely those parts which are left at Kirkdale and devouring what are there wanting . . . So wonderfully alike were these bones in their fracture . . . that it is impossible to say which bone had been cracked by Billy and which by the hyenas of Kirkdale!'

Caricature of William Buckland entering Kirkdale Cavern
by William Conybeare.

Buckland gathered more than three hundred hyena canine teeth from the cavern, and the bones of over seventy-five hyenas. Comparing these to skeletons of living species, Cuvier showed 'that the fossil hyena was nearly one third larger than the largest of modern species. Its muzzle was shorter and stronger . . . and its bite more powerful.' Since it was a species of hyena from genera that now only inhabit the tropics, Buckland reasoned that there had once been a tropical climate in Northern Europe. His interpretation of the cave as an ancient hyena den has proved correct, and when he presented his ideas to the Royal Society they were so well received that he was honoured with the Society's prestigious Copley medal, never before given to a geologist.

Buckland told the Royal Society that the hyenas thrived in the

'Ante-diluvian period, immediately preceding the Deluge', and speculated that the extinct species in the cave were destroyed during the
biblical Flood. These conclusions were based on the supposition that
there were no human records of the species living in Europe since the
Flood. As the bones were so well preserved in mud and silt he maintained the animals had been destroyed suddenly, and from the quantity
of stalagmite in the cave above the mud he estimated that the inundation
occurred six thousand years ago. In 1823, Buckland published a full-scale
treatise, the '*Reliquiae Diluvianae*, or Relics of the Deluge', in which he
tried to fit this cave study and his earlier work on gravels with Cuvier's
most recent 'catastrophe'.

Cuvier's studies in the Paris basin had suggested that during each local
catastrophe the land and the sea had changed places; this was reflected in
alternating layers of marine and land strata. Buckland maintained that
since the Yorkshire cave was inhabited by hyenas *before* the catastrophe
that destroyed them, the area was land both before and after the Flood.
The Flood, he reasoned, had been a transitory event during which the
land remained in the same position. This lent weight to his view that any
Flood should be viewed as a surge or tidal wave rather than a prolonged
event. He also tried to show that the Flood had covered the whole globe.
The fossils retrieved from the caves were identical to fossils found in
loam and gravel deposits all over Europe, and so Buckland speculated
that the same catastrophic event had destroyed the animals in the cave
and swept the gravels to their positions. The gravel deposits were found
in similar circumstances all over Europe, including hill sites, 'to which
no rivers could ever have drifted them'.

Although *Reliquiae Diluvianae* was immensely popular and sold out
almost immediately, it unleashed a storm of comments from literalist
theologians who believed in sticking to the letter of the Bible and disliked
any conclusion that appeared to reduce the power of the Deluge. Rather
than the caves being hyena dens, argued the Reverend George Young, a
minister from Yorkshire, the awesome violence of the Flood had torn
animals apart, limb from limb, forcing the confused debris of many of
them into fissures in rocks and caves. The fractures and 'bite-marks'

were not due to their having been eaten, but rather, testimony to the 'wild confusion' of the torrent in which the creatures were tossed and mangled. Others too, disputed that tropical animals had once lived in England. Tropical beasts were found in Yorkshire because the mighty currents had swept them thousands of miles. 'Can we conclude with geologists that England must once have been inhabited by tropical animals merely because their remains are now found there, in a scattered and broken state?' protested the theologian George Fairholme. 'Had this not been the hypothesis of some of our ablest geologists it would have been termed the result of the most inconsiderate ignorance!'

As a backlash developed in response to Buckland's interpretation of the Flood, other theological scholars challenged the idea that the Flood affected only the surface of the globe. In Moses' account, 'all the fountains of the deep' were opened and the earth's crust was totally destroyed by a mighty, raging torrent. According to Buckland, the Flood was a rather more modest affair, merely confined to shifting the superficial gravels. It wasn't long before literalists objected to Buckland's fundamental premise that geological epochs of immense duration had occurred before the Flood.

Layers of rock thousands of feet thick were demolished during the Deluge, according to the biblical scholar George Cumberland. 'The fountains of waters contained in the great depths of the earth were broken up,' he said. 'Universal subsidence must have taken place. The operation must have been pretty rapid and immense layers of strata must have formed, filled up with the debris of the broken surface.' Far from strata forming almost imperceptibly over countless years, there was a 'sudden production of a thick sequence of rock!' he claimed. 'Such a world as ours might very well come forth in all its finished beauty instantaneously.' The Reverend Young even produced an estimate of the speed of formation of the earth's crust: 'Provided there are currents to supply the materials, strata can form at a rate of nine hundred feet in a month!' he declared.

George Fairholme captured the sense of outrage at the insolent new science that dared to challenge biblical records: 'It is not unknown what ungodly avidity is exhibited by infidel philosophers . . . to distort every fact of science into a sophism against the Scriptures of eternal truth. Of these open scoffers . . . we have no dread; for the Bible has nothing to lose by being tried, like gold in the hottest crucible,' he preached. 'The gates of Hell itself cannot prevail against the word of God.'

William Buckland, with his blustering self-confidence and tremendous enthusiasm for his 'noble subterranean science', tried, as usual, to steer a path through these obstacles. But even his colleagues at the Geological Society questioned some of his evidence. How could he assume that the Flood was global, when gravels were found only in northern latitudes? The more the Reverend Buckland struggled to fit the findings of geology with the Bible, the more anomalies seemed to arise. Was Noah's Flood transient or prolonged, global or local? Did the waters destroy only superficial layers or the entire earth's crust? Were animals made extinct in one biblical Flood, or in a series of Cuvierian 'catastrophes'? Or even, as Lamarck proposed, were species not truly extinct at all, merely transmuted into other creatures.

With some justification, one Scottish minister, John Flemming, summed up the confusion in a paper in the *Edinburgh New Philosophical Journal*: 'The Geological Deluge, as interpreted by Baron Cuvier and Professor Buckland, [is] inconsistent with the testimony of Moses and the Phenomena of Nature.' In Oxford, Buckland's dilemmas were immortalised in a popular satire, *Facetiae Diluvianae*, in which Buckland met the great prophet Noah and each added to the bewilderment of the other.

Caught up in the storm at the birth of the new science, it is hardly surprising that the beleaguered Professor Buckland failed to announce the improbable discovery of a forty-foot reptile. However, Georges Cuvier in Paris was getting impatient since he wished to incorporate the information on the Stonesfield reptile in the updated volumes of his *Recherches sur les Ossemens Fossiles*. In September 1820, his assistant Joseph Pentland wrote to Buckland from the Muséum National in Paris: 'Will you send your Stonesfield reptile, or will you publish it yourself?' Deeply immersed in controversy, Buckland hesitated. A year later, the

Reverend Conybeare also referred to the giant carnivorous lizard of Stonesfield in his paper on the *Ichthyosaurus*, adding 'it is hoped [that Buckland] may soon communicate the results of his observations to the public'. But he did not. Soon, Pentland wrote once more, urging Buckland to announce the details of his research. Yet again, Buckland did nothing.

Thus the enormous bones continued to lie in the Ashmolean Museum, carefully prepared and neatly displayed behind the glass cages, an unexplained curiosity. They had become almost invisible by long acceptance, for over a century part of the paraphernalia of the museum alongside the stuffed animals and other objects. For the time being, in Oxford, the question mark they posed over the nature of giant reptilian beasts that had once lived on land was carefully and assiduously not seen.

4

The Subterranean Forest

To see a World in a Grain of Sand
And a Heaven in a Wild Flower,
Hold Infinity in the palm of your hand,
And Eternity in an hour.

William Blake, 'Auguries of Innocence'

While William Buckland was preoccupied with grand theories and finding little time to investigate the giant reptile of Stonesfield, Gideon Mantell was rapidly becoming obsessed with the strange fossils emerging from the Weald in Sussex. As he began to prepare his first book, *Fossils of the South Downs*, during the late autumn of 1821, he wrote, with some excitement, that 'the relics of a former creation' that he had uncovered were as 'extraordinary as any hitherto recorded'.

Everything about this secret, hidden world, buried beneath the Sussex landscape, seemed bizarre and unpredictable. One persistent puzzle was why the bones of large reptilian creatures should be found with fragments of tropical vegetation? After his first discovery in 1820 of what appeared to be an ancient 'palm' entombed in the quarries at Whiteman's Green, Gideon Mantell tried to find out about tropical botany through his contact Charles Konig, at the British Museum.

Tropical plants had been known in Britain since Captain Cook, having discovered the east coast of Australia, Java, and Easter Island, returned from his voyage on the *Endeavour* in 1771. Accompanied by the botanist Joseph Banks, Cook had brought back hundreds of specimens that he had donated to the British Museum. Banks had later persuaded George III to turn Kew Gardens into a botanical research centre, displaying plants from all over the world. From these eighteenth-century explorations the English horticulturalists began to learn more about the hot, wet eco-systems, unmarked by seasons, within which these plants flourished.

Gideon Mantell set about tracing specialist sources of living tropical plants in order to compare the fossils he uncovered. He was 'much pleased' with 'the unrivalled collection of living palms of Messrs Loddiges of Hackney', one of the few palm merchants in Georgian Britain. As news of Mantell's curious finds spread, local people, too, provided unexpected help, such as: 'the Honourable Mrs Thomas of Ratton, Eastbourne, who presented interesting specimens of the trunks of fossil palms from Antigua'. From these comparisons, Mantell deduced that several of the fossil stems and trunks he was uncovering with the giant animal bones were from ancient tree-ferns. 'The surface of these fossils is rough, the trunk is nearly cylindrical . . . They resemble species of arborescent fern, perhaps Dicksonia?' he speculated. *Dicksonia* is a contemporary tree-fern that can reach a large size, with a slender stem and huge fronds. Mantell sent fossils to Konig at the British Museum, who confirmed his suspicions: 'Some tree ferns are very like this with regard to the lozenge-shaped bases of the fronds,' he replied.

The largest fossil trunk in Mantell's collection was fourteen inches in circumference and four feet in length. From the thickness of this trunk and the rudimentary branches it looked as if it had once extended a great deal further and was part of something tall and tree-like, not a little shrub. Mantell compared the measurements of this trunk to those of tree-ferns in New South Wales, which could grow to thirty feet with stems of only a foot in diameter. 'From the imperfect state in which these [fossils] occur it is evident that the originals attained a very large size,' he wrote incredulously. Huge tropical plants alongside huge reptilian animals: it was barely believable.

Yet each trip to Loddiges' Greenhouses provided more evidence. Mantell soon identified cycads: 'the impressions of the leaf stalks on the bark bear a great resemblance to those on the stems of Cycas revoluta,' he wrote. Cycads look similar to short palms, the trunk covered with the woody bases of leaf stalks and bearing a big crown of leaves at the top. There were also fragments of unknown foliage, heavily blackened with charcoal and quite unlike anything in Loddiges' Greenhouses. 'These specimens are so entirely distinct from any that are known to exist in European countries that we seek in vain for anything analogous,' Mantell observed. Many of the fossils he uncovered are now known to have been *Bennettitales*, an extinct group of cycad-like plants once dominant in the ancient Weald.

Concealed with this buried tropical forest were the remains of aquatic invertebrates. From his early studies on the Downs, Gideon Mantell was an expert on the marine invertebrates of the chalk deposits. The invertebrates of the Weald were different. He could not see the familiar whorls of the ammonite or snake-stone, of belemnites, nautilus or other shelled creatures which once swarmed in the primitive seas that formed the chalk. Instead there were the casts of shells that he did not recognise; impressions sometimes so faint that they left just the barest trace of their external forms: the hinge of two joined shells, as in certain types of clam and pearl mussel, or the fragmentary pieces of a species of snail, perhaps. It was indeed tantalising; fragments both familiar and unfamiliar, never quite forming a complete fossil or displaying a clear marking. Uncertain what they could be, Mantell wrote to his usual correspondents such as James Sowerby, an expert on fossil shells, hoping he would shed more light on these invertebrates.

As for the massive animal bones that were scattered among the debris of this tropical forest, they remained indecipherable; an ancient hieroglyphic for which he did not have the code. He was increasingly certain that many of the bones, such as the giant thigh bone, did not match those of the sea lizards. They were far too chunky and solid. Although some of the bones were rather like those of ancient crocodiles, he had two sets of very large teeth that were not: the worn teeth of a herbivore and the blade-like teeth of a carnivore. 'Of the numerous specimens in my collection not one is perfect; by far the greater part consisting of fragments rounded by the action of water and deprived of the anatomical distinctions so necessary to the elucidation of the form

The quarry at Whiteman's Green in the Tilgate forest where Mantell, shown sitting in the foreground, found large bones of ancient reptiles.

of the original,' he wrote, utterly baffled by these remnants of a 'former creation'.

His investigations were becoming so compelling that other aspects of his life paled by comparison. 'Murdered two evenings at cards,' he complained in his diary. Whether attending the local sheep fair or the ever-popular Brighton races, as a doctor he had a position to maintain in the heart of the community. In provincial society it wouldn't do to appear hurried, or unavailable. But each night when his duties were done he would pore over the details of the animal bones and tropical vegetation, trying to make sense of the wild profusion of relics from this ancient time.

On the evening of 4 October 1821, an unexpected visitor arrived at Castle Place who was able to help him. Mantell was summoned downstairs to meet a young man who 'presents nothing remarkable, except a broad expanse of forehead,' he wrote. 'He is of the middle size . . . small eyes, fine chin and a rather reserved expression of countenance.' The stranger introduced himself as Charles Lyell. Lyell had been visiting his former school in Midhurst, Sussex, when quarrymen had told him of a 'monstrous clever mon, as lived in Lewes . . . who got curiosities out of the chalk-pits to make physic with'. The quarrymen were Mantell's labourers, and Lyell was so intrigued by their account that he rode for twenty-five miles across the Downs to track the man down.

It was soon apparent that Lyell and Mantell had a great deal in common. 'Mr Lyell is enthusiastically devoted to geology,' Mantell entered in his diary; 'he drank tea with us and we sat chatting on geological matters till now – midnight'. Lyell's interest in geology had started while at Oxford University. Although studying classics, he had been drawn to Buckland's inaugural lectures in which the professor was at his most electrifying. Lyell's father had written to a friend, 'Buckland's lectures are engaging [my son] heart and soul at present.' Afterwards, in keeping with his position as the eldest son of minor gentry, Lyell had embarked on a career in law in London, but his eyes gave him trouble. Eventually, his father had indulged his interest in science and taken him to Europe. During one carriage tour across the Alps, Lyell had studied the effects of glaciers on the landscape; on a second trip, he had observed the effect of rivers in forming a coastal plain on the Adriatic coast of Italy.

Since his family was wealthy, with a large estate in Scotland, Lyell had an independent income and more leisure for geology than Mantell. The following day, while Mantell was visiting patients, he went to explore the Sussex strata and then returned to Castle Place: 'to have tea at six o'clock,' Mantell wrote. 'My few drawers of fossils were soon looked over, but we were in gossip until morning.' The visit marked the beginning of an enduring friendship between these two men, both hoping to make a career from geology.

Although there is no record of their conversation over these two days, there is evidence that Lyell told Mantell of Buckland's giant reptile in the Ashmolean Museum and they compared the Stonesfield fossils in Oxfordshire with those of Cuckfield in Sussex. Fired by these discussions, soon after leaving, Lyell lost no time in visiting Stonesfield to obtain a boxful of fossils that he despatched to the Lewes wagon office. Three weeks later, on 25 October 1821, Mantell wrote in his diary: 'received an interesting collection of Stonesfield fossils from Mr Lyell; in many respects they resemble those of Cuckfield'.

Charles Lyell's news of the huge reptilian bones in Oxford confirmed for Mantell that his fossils were not just of provincial interest. He learned not only that Georges Cuvier had concluded that the Stonesfield beast was a reptile, but also that it was at least forty feet long and as bulky as an elephant. Armed with this information, Mantell felt that his own speculations of giant lizards buried in the Weald did not seem quite so preposterous. He could now attempt to classify his own fossils by seeing which bore most resemblance to the giant Oxford lizard.

About this time, Mantell almost certainly heard from Lyell of William Buckland's intention to publish a detailed paper on the Stonesfield reptile. Since Buckland, the famous Regis Professor, was planning to describe and name the new carnivorous lizard, it was hardly appropriate for the unknown Mantell to claim this opportunity for himself. However, no one had reported anything like the unidentified herbivorous teeth. Mantell felt, therefore, that he could be the first to identify this animal, new to science, and claim the recognition, without interfering in Buckland's study.

Patiently taking advantage of any introduction he could negotiate, Gideon Mantell sent a prospectus of his planned book on the geology of Sussex to members of the landed gentry, inviting them to subscribe for copies. The Earl of Chichester, the Bishop of Durham, the Earl of Egremont and numerous others replied; in all he attracted two hundred subscribers. Better still, in 1821 an envelope arrived from Carlton House Palace. Mantell broke the royal seal, and read: 'His Majesty is pleased to command that his name should be placed at the head of the subscription list for four copies.' Quite how George IV had heard of the book is unclear; Mantell wrote back simply, 'I am indebted to J. Martin Cripps Esquire for this honour.' But there can be no doubt of Mantell's response: the royal encouragement was, he said, 'most gratifying to my feelings'. He had great expectations now that his book would place him 'in the first circles' and allow him some means of devoting more time to geology. The carelessly rich could so easily liberate him from his unrelenting daily round of chores.

Fossils of the South Downs, published in May 1822, reveals the progress Gideon Mantell had made in interpreting the strange fossils buried in the Weald. In the preface he pointed out 'that his labours were snatched from hours of repose . . . a record made under circumstances unfavourable to literary pursuits', and he even apologised for the quality of his wife's drawings. 'As the engravings are the first performances of a lady but little skilled in the art, I am most anxious to claim for them every indulgence . . . although they may be destitute of that neatness and uniformity which distinguish the works of the professed artist, they will not, I trust be found deficient in the more essential requisite of correctness.'

Gideon Mantell began by classifying the strata of Sussex. The lowermost and oldest Secondary rock he identified as the 'Iron Sand'. Above this in order of succession he placed the limestone, sandstone and slate where he had found the giant bones, calling this the 'Tilgate Beds' named after the Tilgate Forest. This was followed by Weald clay, greensand and several chalk formations. On top of these Secondary layers came the more recent Tertiary formations such as London clay. He described many of the fossils he had found in the chalk. At a time when palaeoichthyology, the study of fossil fish, was unknown, Mantell had collected superb fish specimens. He also classified fossil invertebrates of the chalk and named more than sixty new species, including different types of ammonites, zoophytes, echinites, univalves and bivalves.

With some understatement that belied the months of feverish excitement, Gideon Mantell stated that the Tilgate beds in the Weald were 'one of the most important series of deposits' that he had uncovered. He attempted to catalogue the extraordinary fossils of the giant bones. Under the heading 'Fossil Lacertae [Lizards]' he wrote: 'the teeth, vertebrae, bones and other remains of an animal of the Lizard Tribe of enormous magnitude are perhaps the most interesting fossils that have been discovered in the County of Sussex'. He described the characteristics of the sharp, curved carnivorous teeth and provided measurements of fragments of vertebrae and ribs, which were, he said, 'decidedly analogous to those of the Lizard Tribe'. Other bones were also listed: the head of the radius (forearm), metacarpals (bones of the hand) and a thigh bone. 'Some fragments of a cylindrical bone, probably the femur, indicate an animal of gigantic magnitude,' he observed. 'I have specimens from ten to twenty-seven inches long and from eleven to twenty-five inches in circumference, the substance of the bone being more than two inches thick.'

Recognising from the herbivorous teeth that he had evidence of a second type of giant creature different from the carnivorous Oxford monster, but perhaps not liking to court controversy by suggesting he had found a herbivorous lizard, he classified other giant bones under a different heading: 'Teeth and Bones of Unknown Animals'. He wrote: 'a brief description of these fossils is here inserted not in the hope of being able to elucidate their nature, but to record their existence in the Tilgate Forest with a view to future enquiries . . . [The teeth] are of a very singular character and differ from any previously known.' He had the crown of the teeth only, he explained, unattached to the jaw. Although they were worn, some specimens were 1.4 inches long: 'when perfect these specimens must have been of a very considerable size'.

Mantell even pointed out the analogy between the fossils of Tilgate and those of Stonesfield in Oxfordshire. Perhaps in a gentle spur to Professor Buckland, he wrote 'the Stonesfield limestone has long been celebrated for the extraordinary character of its fossils, of which however, no detailed account has yet appeared before the public'. With the assistance of Mr Charles Lyell 'and aided by an interesting collection of Stonesfield fossils for which I am indebted to his liberality,' he continued, 'I have been able to ascertain that the following organic remains occur in both deposits:

The teeth, ribs, and vertebrae of a gigantic animal of the Lizard Tribe.
Bones and plates of several species of Tortoise.
Teeth of a species of *Anarhicas* [wolf-fish].
Scales of Fishes and Lizards.
Bones of Birds? and of Quadrupeds [unknown]'

In his conclusion, Mantell stated boldly that entombed in the hills of Sussex 'there are one or more gigantic animals of the Lizard Tribe'.

Although he could not name the creatures or have any clear conception of the kind of beast he was describing, this was the first attempted scientific description of dinosaur remains correctly identified as giant lizards. It was a vivid snapshot of a wondrous unknown past. 'We know not the millionth part of the wonders of this beautiful world,' he wrote. 'It is the pleasing task of the geological inquirer . . . to discover order and intelligence in scenes of apparent wildness and confusion . . . to recognise a series of awful but necessary operations by which the harmony, beauty and integrity of the universe are maintained . . . which must be regarded as wise provisions of the Supreme Cause.'

As he proudly received the first printed copy at the beginning of May 1822, he had high hopes that this would prove a turning-point in his career. 'I am resolved to make every possible effort to obtain that rank in society to which I feel I am entitled both by my education and my profession,' he wrote in his journal. Surely, fired by these strange findings, some rich patron would step forward; his endless round of medical duties that took up so much of his time would, perhaps, soon be a thing of the past? At the very least, he hoped that his labours would be well received by the prestigious London societies: the Royal Society and the Geological Society.

Soon after the publication of his book, Mantell took some of his Sussex fossils to a meeting of the Geological Society in Covent Garden. The worn teeth of the giant herbivore were carefully wrapped in cloth. It was a long and tiring journey to London by chaise, stopping several times to change the horses, before he found his way to the House of the Geological Society at 20 Bedford Street. The Reverend William Buckland, now the Vice-President of the Society, had come down from Oxford with his friend the Reverend Conybeare. William Clift, Conservator of the Hunterian Museum of the Royal College of Surgeons, was also present.

Mantell's diary and his subsequent accounts reveal that after the business of the meeting was completed, he showed these experts some of the worn, brown teeth from his unknown herbivorous animal. 'I was discouraged by the remark that the teeth were of no particular interest,' he wrote. The experts did not agree with Mantell that the 'tooth' belonged to an ancient herbivorous lizard. Far from such an exotic and fanciful verdict, they claimed: 'There is little doubt the teeth belonged either to some large fish, allied to "Anarhic[h]as lupus" or wolf-fish, the crowns of whose incisors are of a prismatic form, or were mammalian teeth obtained from a diluvial [recent] deposit.'

Thus the combined wisdom of these august members of the Geological Society was that the tooth on which Mantell had pinned all his hopes belonged to nothing more exotic than a recent mammal such as a rhinoceros or an oversized fish! Mantell felt their dismissive lack of interest keenly. How could anybody build a reputation on a large fish? There was only one person there who dissented from the expert verdict

— William Hyde Wollaston — and he happened to be the only person present who was not a geologist.

The scepticism of the experts at the Society stemmed from the fact that they did not accept Mantell's classification of the strata of the Weald as Secondary rock. His conclusion that he had found a giant herbivorous lizard could be wrong if his interpretation of the Tilgate beds as ancient Secondary rock was incorrect. Numerous mammalian remains had been found in the more recent Tertiary rocks which lay above the Secondary strata: mammoths, elephants, rhinoceros and hippopotamus. If the Weald rocks in Sussex were Tertiary, then the giant fossils within them, far from belonging to some improbable species of herbivorous reptile, were much more likely to be from any of these large mammals. To persuade the experts that he had indeed found an ancient reptile, he had first of all to prove beyond doubt that the Tilgate beds were Secondary rock.

The eminent members pored over the details of Mantell's findings and tried to fathom whether the limestone and sandstone of the Tilgate Forest were part of the 'Purbeck' formation, or 'Ferruginous sand', 'Greensand', 'Iron-sand' or 'Hastings sand'. Their task was made all the harder since the names for the Sussex strata were not yet standardised and everyone was using different terms for the various layers, adding to the bewilderment. For Mantell, with each learned utterance from the experts the years of painstaking work were falling away, the exotic lizards of mythical proportions fast fading into nothing more than a figment of his imagination. He was just a country doctor, after all.

There was good reason to be confused when trying to place the strata of the Tilgate Forest into the geological sequence. Unlike the Stonesfield rock near Oxford where the fossils were found deeply buried, the rock at Whiteman's Green in the Weald was inexplicably close to the surface. Was this, as Mantell claimed, a protrusion of older, Secondary rocks? Or was it a recent deposit, perhaps of Tertiary or even younger alluvial rocks, as Buckland thought. In *Fossils of the South Downs* Mantell made no attempt to conceal his perplexity about the exact position of the strata in which he had found his giant reptiles. Although

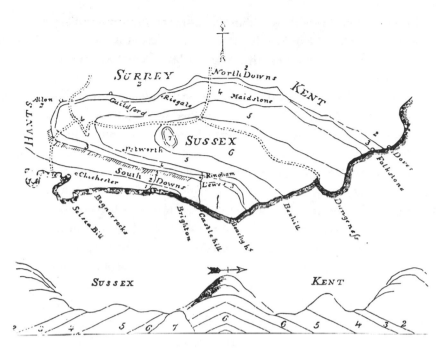

The stratigraphic sequence in Sussex as drawn by Mantell in *The Fossils of the South Downs*, 1822. Were the rocks of the Weald [no. 7] a protrusion of older rocks to the surface or a recent deposit of younger rock?

he had correctly identified the Tilgate Beds as Secondary, he did admit that the precise 'geological position of these beds [within the Secondary series] is involved in much obscurity and cannot at present be satis-factorily determined'.

Faced with the disbelief of the Geological Society, shortly after this meeting Mantell made yet another survey of the Sussex rocks, this time with his friend Charles Lyell. Riding west from the Tilgate Forest, Lyell and Mantell searched for quarries that contained strata and fossils that matched those found at Whiteman's Green. They were hoping to find a site where the different layers of rock were clearly exposed in the geological sequence, so they could prove beyond doubt the exact position of the Tilgate Beds within the Secondary series of rocks. If they could convince the experts that the Tilgate rock was Secondary, then surely no one would doubt that Mantell had indeed found an ancient giant lizard?

To Mantell's delight, they uncovered similar organic remains — bones, teeth and 'numerous vegetables allied to the Cycas' — in the sandstone cliffs of Hastings, Rye and Winchelsea. Even better, in a quarry near Rye they found the strata laid bare. Sandstone and limestone matching the Tilgate beds were embedded in the Secondary rock known as Iron-sand.

After this expedition, on 1 June 1822 Mantell wrote triumphantly to Dr William Fitton, the Secretary of the Geological Society: 'I think we may fairly conclude that the sandstone of Rye, Winchelsea, Hastings, Tilgate Forest and Horsham are but different portions of the same series of deposits belonging to the "Iron-sand" formation.' Mantell was now completely satisfied that the limestone and sandstone in which he had found the giant bones in the Weald could be placed in the Secondary series, well below the chalk formations. Consequently, in his letter to the Geological Society he went even further. In defiance of the experts such as Buckland, he restated his own interpretation of the animal remains that he had found. The large herbivorous teeth were now clearly identified as 'Teeth of an unknown Herbivorous Reptile, differing from any hitherto discovered either in a recent or fossil state'. In addition, he confirmed that he had the teeth and bones of a lizard resembling those found at Stonesfield, and 'Teeth and bones of crocodiles and other Saurian [lizard] animals of an enormous magnitude'. From the evidence of this letter the amateur Gideon Mantell was in no doubt that his beguiling view of a buried ancient world inhabited by several different species of giant reptiles — herbivores *and* carnivores — was an accurate one.

However, his letter was regarded as of such insignificance by senior members of the Society that it was not even read out, as planned, to the eminent company. For one thing, George Bellas Greenough, a Fellow of the Royal Society, a former MP and the first Chairman and President of the Geological Society, was convinced that iron-sand was always a *marine* deposit. Since Mantell had reported some *freshwater* shells mixed in with the giant bones, Greenough insisted that the Tilgate beds could not be iron-sand and refused to change his opinion in the light of Mantell's findings.

William Buckland, too, was certain that the Weald rock resembled a recent, Tertiary rock he had seen while travelling in Italy and so was not a Secondary stratum. Consequently, in his view, Mantell's 'reptiles' had to be large mammals. And such was the standing of both Buckland and Greenough that other members could not accept that a provincial surgeon could possibly have knowledge that surpassed that of the Oxford and London men who were the leaders in the field.

Over six months elapsed before it was decided that Mantell's letter to Fitton on the strata of the Tilgate Forest would be read before the Geological Society. The minutes of the meeting on 17 January 1823, show that both Lyell and Mantell were present. At the Council committee meeting the following week, Mantell's paper was read and passed on to referees to check before publication. However, it remained unpublished for a further three years. The archives reveal that Gideon Mantell had considerable difficulty getting his papers published by the Society. One unsigned letter from a referee considering his paper on fossil vegetables wrote: 'the notice is not of sufficient importance to be printed'. George Greenough, too, turned down Mantell's paper on the Tilgate Forest. William Buckland was so convinced that Mantell was wrong, he wrote specifically to warn him against claiming that the teeth and bones were found in 'the older Iron-sand formation'. Mantell believed this advice came from the best of intentions and commented on 'the generous kindness that marked his character'.

Mantell's uphill struggle to get his ideas accepted by the experts was not unique. One amateur geologist, Robert Bakewell, who was not allowed to join the Geological Society although he wrote a popular book, *Introduction to Geology*, wrote frankly about the difficulties. 'There is a certain prejudice,' he said, 'among the members of the Scientific Societies in London and Paris, which makes them unwilling to believe that persons residing in provincial towns or the country can do anything important for science.' William Smith, the surveyor who pioneered studies of strata in England and was also not a member, once remarked: 'the theory of geology was in possession of one class of men [at the Geological Society] and the practice in another'. Gideon Mantell, an amateur from the provinces with none of the trappings of the upper classes, was very much an outsider. The disappointment he felt at the rejection of his ideas, and his failure to obtain recognition for his giant lizards, was recorded in his diary:

> The past year, like its predecessors, has fleeted away almost imperceptibly, and I am as far from attaining that eminence in my profession to which I aspire as at the commencement of it. The publication of my work on the Geology of Sussex . . . has not yet procured me that introduction to the first circles . . . which I had been led to expect it would have done. In fact I perceive so many chances against my surmounting the prejudice which the humble situation of my family naturally excites in the mind of the great, that I have serious thoughts of trying my fortune in Brighton or London.

Since in the eyes of the experts at the Geological Society there was so much doubt about the stratigraphy of the Weald – and this was crucial for interpreting the animal remains – the Secretary William Fitton came down from London to settle the matter once and for all. Fitton, unlike Mantell, was a veteran of two universities, having studied at both at Edinburgh and Cambridge. To add to his flawless academic pedigree, he was also a Fellow of the Royal Society. His good fortune had been assured when in 1820 he married 'a most amiable lady, who brought him the means of a comfortable existence'. So comfortable, in fact, that he was able to retire and devote himself to geological pursuits.

With Gideon Mantell providing much of the fossil evidence, Fitton started his investigations of the Weald in 1822. Since Mantell was invariably busy with his medical duties he was often unable to accompany the geologist on his forays around Sussex, but gradually Fitton began to make sense of the rocks of the Weald.

The first clue to the origin of the Weald strata came from the pronounced 'ripple marks' in the sandstone in the Tilgate Forest. These marks resembled the pattern of sand on a beach, as though countless shallow wavelets had moved across the soft strata. Mantell had already noted the 'extraordinary appearance' of this rock, 'being everywhere marked with undulating furrows, so strikingly resembling the impressions made on the sand . . . by the action of waves'. Dr Fitton speculated that the rocks might have been formed in a low-lying flood plain or along the edge of a lake or river delta, where sandbanks had accumulated.

A careful study of the shells embedded in the rock confirmed that the Tilgate beds were indeed part of a freshwater deposit. William Fitton took some of the shells to Paris to discuss them with invertebrate specialists such as Adolphe Brongniart, the son of the distinguished Alexandre Brongniart, the colleague of Cuvier. From the scanty impressions of the creatures Fitton was able to identify nine or ten species of univalves and bivalves. Some of these molluscs, such as the *Unio valdenisis*, a clam or pearl mussel, could never have survived in salt-water. Fitton also realised that the Sussex marble embedded in the Weald clay, used for centuries to adorn the walls of Petworth Priory and other grand local houses, was formed from the shell of another freshwater creature: the *Paludina* snail. Gradually Fitton and Mantell became experts on freshwater shells such as *Planorbis*, *Lymnia*, *Paludina* and *Cyrena*. Fitton speculated that the freshwater deposits of the Weald could be explained if this region had once been a vast river delta.

But if the Tilgate beds were a freshwater deposit, how could they be part of the marine iron-sand in the Secondary series? It took Fitton several years to make sense of his data. Finally, he realised that the assumption made by Greenough and other senior members of the Geological Society that the iron-sand could only be marine was wrong. There were *two* types of iron-sand, one marine and the other freshwater. 'And now all things fell into their proper places,' Fitton wrote to a friend. The Tilgate beds were part of the freshwater iron-sand. And as Mantell had claimed all along, these rocks were indeed Secondary.

Fitton's confirmation that the rocks of the Tilgate Forest were freshwater deposits made sense of the animal data too. Creatures like crocodiles would not be found in the deep seas, as would ichthyosaurs, but would lurk in rivers or flood waters. 'In fact the existence of dry land at no great distance seems clearly indicated by these remains of vegetables and amphibia; some of the former must have grown on the borders of a river or lake,' Mantell wrote. 'At what period was it and under what circumstances that turtles and gigantic crocodiles lived in our climate and were shaded by forests of palms and arborescent ferns?'

The more that was revealed of this vast buried river delta, the more this helped to confirm Mantell's understanding of the animals that had lived close by. This was not the ocean world of Lyme Regis that Mary Anning was uncovering. What they were beginning to catch glimpses of here was a Cretaceous landscape: 'a mighty river flowing in a tropical climate over sandstone rocks . . . through a country clothed with palms and arborescent ferns . . . inhabited by turtles, crocodiles and other amphibious reptiles'. A scene where the uppermost fronds of a forest of giant tree-ferns emerged above the mists that wreathed the river delta, and tropical plants provided plentiful vegetation for giant herbivores to graze. As Mantell worked in the quarry, occasionally catching glimpses of the English country scene beyond, it was hard to believe that beneath the order of thousands of years of careful cultivation were the long-buried relics of such an alien landscape.

Unravelling the layers of rock that formed the Weald proved so complex that it took Fitton several years before he could confirm his finds, and the results of his investigations were not published in full until 1833. However, ten years before this, Charles Lyell also obtained conclusive proof that the Weald was indeed Secondary rock.

In June 1823, Lyell made an expedition to the Isle of Wight with Professor Buckland and other geologists. Along the southern side of the island between Compton Chine and Brook the strata are clearly exposed. All the Sussex rocks that Mantell was struggling to place in order are present, so it was possible to see at once which beds are above and below the others. Thus on 11 June Lyell wrote excitedly to Mantell: 'we see there, at one view, the whole geology of your part of the world, from the chalk with flints down to the Battle [the Sussex town] beds, all within an hour's walk, and yet neither are any of the beds absent . . . This is so beautiful a key that *I should have been at a loss to conceive of how so much blundering could have arisen if I had not witnessed the hurried manner in which Buckland galloped over the ground.*'

Lyell stayed a day later than the rest of the party to gather fossils, and wrote disparagingly of the 'confusion which has found its way into the heads of some of our geologists with regard to your Sussex beds'. His beautiful cliffs did indeed prove the position of the Weald rocks in the Secondary series. It restored Mantell's conviction that the fossils he had found could actually belong to an ancient giant reptile.

But who would believe them? Buckland's view that the Tilgate beds of the Weald were recent had become the established view. Charles Lyell, not yet twenty-five, had graduated in geology only three years previously and was still a law student. Gideon Mantell, shoemaker's son, provincial doctor and part-time geologist, scarcely carried more weight. The Royal Society turned down Mantell's first application, in 1823, to become a Fellow. The Geological Society still would not publish their evidence that the Tilgate Forest formed part of the older, Secondary strata. William Fitton, with admirable scientific caution, wanted to be sure of his data before he announced his findings. Charles Lyell's letter to Gideon Mantell setting out the evidence from the Isle of Wight was not published, and did not come to light for some years.

Mantell had one last hope. Lyell was planning to visit Paris, and he offered to take some of Mantell's fossils to leading thinkers in France. Mantell hit upon the bold idea of showing the mysterious herbivorous 'tooth' to Baron Cuvier. He could not bring himself to accept Buckland's view that this belonged to a mere wolf-fish. Surely the legendary Baron would provide the right answer? Mantell, like all the others, held Georges Cuvier in the highest esteem. The Baron's 'powerful mind and enlightened genius could, like the fabled wand of the sorcerer, cause to pass before us the beings of former ages,' he wrote. Such were his feelings of admiration for Cuvier that, when he had an opportunity to meet him in person ten years later, Mantell described himself as 'trembling with excitement' for the entire meeting.

By 1823, Cuvier had recently been granted the title Grand-Master of the University of Paris. He was usually to be found there at the centre of an admiring group of followers, drawn by his learning and reputation. Even his private study, or 'sanctum sanatorium', was imposing. 'It is truly characteristic of the man,' wrote Lyell. 'It displays that extraordinary power of methodising which is the grand secret of the prodigious feats which he performs annually, without appearing to give himself the least trouble . . . It is a longish room furnished with eleven desks to stand to . . . like a public office for so many clerks. But it is all for the one man, who multiplies himself as author, and admitting no one into this room, moves as the fancy inclines him, from one occupation to another.'

Any visitor hoping to meet Georges Cuvier would try to obtain an invitation to his Saturday evening soirées. These soirées, wrote Lyell, were attended

> [by] the learned, and the talented of every nation, of every age and of each sex. All opinions were received; the more numerous the circle the more delighted was the master of the house to mingle in it, encouraging, amusing, welcoming everybody, paying the utmost respect to those really worthy of distinction. It was at once to see intellect in all its splendour; and the stranger was astonished to find himself conversing, without restraint, without ceremony, in the presence of the leading stars of Europe: princes, peers, diplomatists, and the worthy savant himself.

It was in this brilliant company that Gideon Mantell's herbivorous tooth was duly unwrapped and presented to the great Baron on Saturday 28 June 1823. According to Sidney Spokes, Mantell's biographer,

Cuvier pronounced his opinion without hesitation. The worn-down 'tooth' was merely the upper incisor of a rhinoceros. When Lyell persisted and presented some metacarpal bones – small bones forming part of the hand of the forelimb – these too were dismissed as from a species of hippopotamus.

Whether Georges Cuvier was correctly informed at this meeting about the uncertainties over the Sussex strata has been a matter of conjecture for science historians. It seems most likely that he had been told by William Buckland that the Sussex rock was recent, and armed with this erroneous information had arrived at the logical conclusion, given the very worn state of the tooth, that it belonged to a herbivorous mammal. There is also some evidence that Cuvier may have modified his opinion the next day, for Lyell later recorded that although he pronounced it to be the incisor of a rhinoceros, 'this was however at an evening party. The next morning he told me he was satisfied it was something quite different.' Strangely, this all-important qualification does not appear to have reached Gideon Mantell. In England, correspondence between Buckland and other scientists reveals that it was now widely accepted that Mantell had merely found a rhinoceros.

On receiving the letter from Paris informing him of Cuvier's verdict, Mantell finally had to accept that it seemed he had been wrong. Quietly reading and rereading it for any sign of encouragement, he could not escape the fact that the letter pronounced a crushing blow, and one from which he could not easily recover. It was clear he had been wasting his time; perhaps he had even made a fool of himself. Far from making a discovery that would turn the scientific world upside down, he had uncovered nothing more than an ordinary modern mammal. He had to accept that his strenuous efforts to understand the former worlds of Sussex were leading nowhere.

His wife, too, felt the burden of his continual disappointments. For seven years, now, she had been accustomed to him returning late after visiting quarries when his medical duties were done, and waking alone in the mornings, knowing he had left before dawn. Hard-won funds were used to pay the quarrymen for new fossils – a sacrifice she was expected to share willingly with him. Even his book was published at a loss, incurring a debt of over £300 largely underwritten by her successful elder brother, George Woodhouse.

For Mary Mantell all that seemed to emerge from this expensive hobby were yet more bones. They were everywhere – row upon row of neatly labelled bones that must not be moved, must not be touched, each proclaiming undisputed ownership of the front room where she should have held sway, perhaps entertaining guests for afternoon tea. Her drawing-room was furnished with dreams, but they were no longer *her* dreams.

So Gideon Mantell found he could no longer always count on the support of his wife. The combination of his domestic circumstances, the endless commitments he faced as a doctor and his failure to gain recognition for his ideas, finally pushed him into a deep depression. His huge energies and boundless ambition became ensnared by an overwhelming sense of frustration and loss. He faced tremendous conflict between his appetite for success, fuelled by the conviction that he had discovered something of tremendous significance, and the apathy with which his discoveries had been received. 'My incessant engagements and occupations [as a doctor] have so constantly engrossed my time that even this journal has been wholly neglected,' he wrote in his diary. 'So unhappily have my days been spent that I had not the resolution to record mementos of wretchedness . . . '

5

The Giant Saurians

Black Ven cliff at Lyme Regis, a dark, forbidding shape visible even through sea mists from the Cobb and the harbour a mile away, continued to surrender to Mary Anning the wreckage of the Jurassic age. After ten years of searching for fossils, every part of the cliff-face had become familiar to her. The grey limestones and shale, the rocks that had precipitated the early death of her father, now provided income for the family. By the early 1820s, Mary had developed such a skilful grasp of where to find fossils that according to one report: 'she would have been able, for instance, out of fifty "nodules" all looking . . . much of a muchness, to pick without hesitation the one which, being cleft with a dextrous blow, should show a perfect fish embedded in what was once soft clay'.

Since Lieutenant-Colonel Birch's sale of his private collection in 1820, which included many of Mary Anning's finds, she had uncovered several different species of ichthyosaurs. In May 1821, accompanied by her dog Tray who would guard her discoveries while she went for help, she excavated the first *Ichthyosaurus platyodon*, a creature almost twenty feet long. Two months later she found a five-foot fossil close by that was named *Ichthyosaurus vulgaris*. Early the next year another large ichthyosaur, this one nine feet long, was retrieved. But despite this success, Mary and Joseph and their mother Molly were often short of money. The fossil-hunting business was erratic, and already by the early 1820s there was increasing competition from other collectors.

So it was with an unaccustomed sense of anticipation that on the evening of 10 December 1823, as she worked at the foot of Black Ven cliff, Mary Anning uncovered something she had not seen before. The object looked like a creature's skull. Yet it was not long and pointed with huge bony cavities for eyes like an ichthyosaur's, but small, a mere four or five inches long, more like that of a turtle. As she excavated around the deeply embedded bones, beneath the circling gulls, it became apparent that the animal had a greatly elongated serpentine frame, with a large number of vertebrae making up the backbone. The neck of the beast appeared to be at least as long as the rest of the body and tail. It was quite dissimilar to any creature yet known to science.

With the help of some villagers, Mary worked through the night against the incoming tide. The conditions were bitter as the winter winds whipped up sand and spray until all were drenched and numb. By early morning, they had gradually revealed the spine of the beast, consisting of ninety bones: the bizarre 'turtle' was nine feet long, more like a snake. They retrieved fourteen ribs and the bones of the pelvis, deeply embedded in shale. Rather than legs like an amphibian's, or fins like a fish's, it had paddles made up of many fine bones.

Mary Anning knew that the experts, such as Buckland's friend Conybeare, had long suspected that in addition to ichthyosaurs a second kind of sea lizard had once roamed the ancient seas. While preparing his study on *Ichthyosaurus* in 1821 with his friend Henry de la Beche, Conybeare had reported large bones that did not match those of the first sea lizard. In Somerset, he had come across a skull that resembled that of a turtle, but was not associated with any shell; paddle bones and un- usually shaped vertebrae added to the mystery. Although he had only a few fossils, the Reverend Conybeare had felt sufficiently confident to suggest a name for the unknown beast: *Plesiosaurus*, meaning 'near to reptile'. Yet even Conybeare had accepted that 'there was reasonable grounds for suspicion' that he had inadvertently created a fictitious animal from the juxtaposition of bones belonging to different species. As Mary Anning surveyed the strange beast before her eyes, she wondered if this could be the entire skeleton of the creature Conybeare had been predicting.

When Conybeare heard the news of her discovery, his peaceful houschold was thrown into such uncharacteristic disarray that he was quite unable to 'settle down to prepare my Sunday sermon'. He wrote a euphoric letter to de la Beche, who was in Jamaica: 'Buckland . . . brought important news – that the Annings had discovered an entire Plesiosaurus and that it had been offered to the Duke of Buckingham for 200 pounds.' Three days later, Conybeare received

a very fair drawing by Miss Anning of the most magnificent specimen . . . It was the evening also of our Philosophical Society at the Bristol Institution and you may imagine the fuss this occasioned. My sermon, though finished in scraps was then not half-transcribed, but one of my sisters-in-law, who was staying with me kindly undertook the task, and to the Society I went . . . Such a communication could not fail to excite great interest; some of the folk ran off instantly to the printing office, whither I was obliged to follow to prevent some strange blunders . . . thus I did not get home till midnight.

But as the news reached Paris, the excitement came to an abrupt end. The eminent Baron Georges Cuvier at the Muséum National d'Histoire Naturelle suspected that the new animal might be a forgery. How could a creature have a neck that was longer than the rest of the body and tail put together? Not even the longest-necked birds, like the swan, have such proportions. The animal deviated in the greatest degree from the almost universal anatomical law that limits the cervical vertebrae, or neck bones, to no more than seven in large quadrupeds. In birds the number of cervical vertebrae can vary between nine and twenty-three. Living reptiles have from three to eight. Yet this allegedly reptilian creature apparently possessed thirty-five vertebrae in the neck alone!

On the basis of drawings sent to him, Georges Cuvier considered it possible that Mary Anning and her family were deceiving the experts. He already had a low opinion of the English anatomists from his dealings with 'the London Baronet', Sir Everard Home, whose early papers on the 'Proteosaurus' had prompted such ridicule. It was easy to imagine that the English experts had been taken in by peasant fossil collectors. The Annings had surely taken the head and neck of a sea snake and attached it to the body of an ichthyosaur? As if to back up Cuvier's point, a convenient crack could be seen in the skeleton at the base of the neck, exactly as if a join had been forged. Cuvier wrote to the Reverend Conybeare warning him to make sure the beast was genuine.

For Mary Anning, Cuvier's suspicion was a disaster. He was the world authority. If he convinced others that the new fossil was a forgery, the Anning family could be ruined. Amateurs had been known to over-restore items: the fossils of the eccentric collector Thomas Hawkins, for instance, occasionally appeared with material strongly resembling Victorian house-brick in them. With the use of lamp-black or soot as a colouring agent, and plaster of Paris for bulk, details were sometimes embellished – a few extra caudal vertebrae, some paddle bones. 'He is such an enthusiast that he makes things as he imagines they ought to be; and not as they really are found,' Mary Anning later observed.

Gossip soon started to circulate about the new beast from Lyme, as letters were hastily penned from one geological enthusiast to another. 'You will much oblige me by calling on Mr Webster [at the Geological Society] who will probably shew you the Original Fossil,' wrote the amateur geologist George Cumberland to a friend in London, 'and tell me your opinion of its being one Fish or not? Ask him his opinion of the New Fish, but do not give him your own.'

A special meeting was convened at the Society to arbitrate on the matter. Mary Anning was not present. Members enquired how any animal could compensate for the weakness that would have attended this great elongation of the neck? Would such a beast even be able to hold its neck upright? Could it move, and if so, how? Assuming its most formidable enemy was *Ichthyosaurus*, with its diminutive head and slender neck it would have been a very unequal combatant, with slow locomotive powers which would make escape by flight almost impossible.

As the gentlemen debated the improbable combination of characteristics of the new creature, it soon emerged that its features matched all the fossil findings that the Reverend Conybeare and de la Beche had uncovered earlier, which had led them, too, to propose the existence of such a beast. 'The magnificent specimen at Lyme,' announced Conybeare, 'has confirmed the justice of my former conclusions in every essential point connected with the organisation of the skeleton.' Others were soon persuaded that Conybeare must be right.

'This really is an extremely curious object,' Charles Konig at the British Museum wrote to a friend. 'It is no doubt perfectly genuine, in spite of the apparent disproportion of the parts, especially the neck and head, which latter is the *smallest* brain-box known in proportion to the bulk of the animal and bespeaks its possessor to have been none of the wisest, though no doubt sufficiently so for his purposes.' As this became the accepted view the Annings were vindicated and, for once, Georges Cuvier was seen to be fallible. Plans were made for the Reverend Conybeare to present a definitive scientific paper on the *Plesiosaurus* at a forthcoming meeting of the Geological Society in early February 1824.

Meanwhile, around the same time during the winter of 1823, *The Gentleman's Magazine* carried an article on some extraordinary discoveries of giant bones in Sussex. Staff at the magazine had picked up a story from the local Brighton papers on Gideon Mantell's collection in which the enormous bones of a carnivorous animal of the 'Lizard Tribe' were clearly described. *The Gentleman's Magazine* was an established national periodical that was posted to many country estates and was certainly available in the Oxford University library for William Buckland to read. After seven years of somewhat leisurely interest in the giant carnivorous reptile bones in the Ashmolean Museum, it seems possible that the Reverend Buckland had suddenly recognised that a little-known, part-time geologist might upstage him. This realisation galvanised him into action and, it appears, temporarily jarred his characteristically generous nature. He wrote to Cuvier in Paris explaining that the remarkable discoveries in Sussex by Mantell made it imperative that he publish details of the huge Oxford reptile soon – and thereby claim the credit for discovering the creature. Buckland decided to announce his Stonesfield reptile to the Geological Society at the same meeting in early February 1824 that Conybeare planned to present his findings on the *Plesiosaurus*. The two men set to work preparing their momentous papers that would surely amaze all of scientific London.

Almost farcically, after all these years of waiting, there was a last-minute hitch. The Reverend Conybeare explained in a letter to Henry de la Beche, who was still overseas, that 'the Duke had shipped the Plesiosaurus to be deposited at the Geological Society where he charged me to meet it on pain of its falling into the hands of Sir Everard Home . . . When I came to town I found the specimen delayed in the Channel. Nor did it arrive for ten days afterwards.' Conybeare's announcement of his findings had to be postponed for a couple of weeks.

Given the enthusiasm with which Conybeare's *Plesiosaurus*, the 'wonder' that was 'coming by water', was due to be embraced by the modern world, there was further embarrassment in store. The specimen had been carefully prepared by the Annings, set in plaster and placed in a huge wooden display case, ten feet tall and six feet wide. This was to be carried upstairs to the meeting-rooms for members to view during Conybeare's talk. But when it finally arrived, despite the sterling efforts of the workmen the massive prehistoric sea lizard could not be manoeuvred beyond the hallway. 'After wasting a day in vainly attempting to move it upstairs to the room of meeting of the Geological Society by the aid of two men, we were constrained to unpack it in the entrance passage,' wrote Conybeare. The gentlemen of the Society were obliged to satisfy their curiosity by peering at the creature in a dark passage by candlelight.

On 20 February 1824, the day scheduled for the first meeting with Professor Buckland as President, the meeting-room was crammed with an eager audience who had heard rumours of the announcements to be

made. Many members had brought visitors: Dr Fitton introduced
Captain Franklin, Charles Lyell brought two friends, a Mr Brookes and
a Mr Hill; George Sowerby was introduced by Mr Children; John
Tilney, a friend of Mantell's, was introduced by Mr Fraser. In all there
were some thirty guests. The usual business of the meeting – thanks,
donations, important notices   was swiftly completed, and then the
Reverend Conybeare was invited to present his 'Notice on the Discovery
of a Perfect Skeleton of the Plesiosaurus'.

Conybeare, for all his intellectual brilliance, was not noted for his
lecturing style. On another occasion, giving a public talk, he is reported
to have 'frightened the ladies' with his 'ungraceful' manner, which
prompted much 'shuffling of feet and rustling of robes'. But it is hard to
imagine that on this particular day he had anything other than the rapt
attention of his enthusiastic audience at the Geological Society.

He described a creature which, even Cuvier now acknowledged, was
altogether the most 'monstrous' that had yet been found amid the ruins
of the former world. After an internment of thousands of years, he said,
the fossil remains were in nearly as perfect a state as the bones of species
that now existed upon the earth. 'To the head of the Lizard, it united the

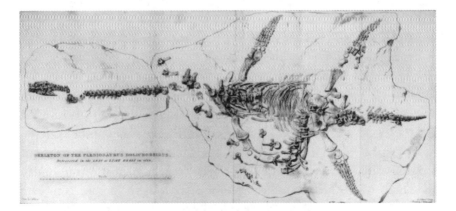

The *Plesiosaurus*, nine feet long, described by the Reverend William
Conybeare in 1824.

teeth of the Crocodile; a neck of enormous length, resembling the body of a Serpent; a trunk and tail having the proportions of an ordinary quadruped, the ribs of a Chameleon, and the paddles of a Whale.' The minutes of the meeting show that Conybeare continued to highlight the distinguishing features of the animal. The head was 'remarkably small, forming less than the thirteenth part of the total length of the skeleton; while in the Ichthyosaurus its proportion is one fourth'. The number of vertebrae exceeded that of any other animal, with a total of ninety joints in the backbone. The paddles 'are composed of two rows of nearly circular bones' which were flattened as in the turtle, leading him to speculate that it moved in a similar way.

'That it was aquatic is evident from the form of its paddles,' he went on. 'However, its long neck must have impeded its progress through the water; presenting a striking contrast to the organization which so admirably fits the Ichthyosaurus to cut through the waves.' Consequently, he reasoned, this giant sea lizard 'swam upon or near the surface, arching back its long neck like the swan, and occasionally darting it down at the fish which happened to float within its reach. It may have lurked in shoal water along the coast, concealed among the sea weed, finding a secure retreat from the assaults of dangerous enemies.' The length and flexibility of its neck might have compensated for its incapacity for swift motion through the water by enabling it to make a sudden and agile attack on any prey which strayed within its extensive sweep. Finally, he concluded by naming the species: in view of its great size it was christened '*Plesiosaurus giganteus*'.

Now William Buckland rose to astound the audience with evidence of an even more gigantic beast. He began: 'I am induced to lay before the Geological Society the representations of various portions of the skeleton of the fossil animal discovered at Stonesfield, in the hope that such persons as possess other parts of this extraordinary reptile may also transmit to the Society such further information as may lead to a more complete restoration of its osteology.' He explained that the evidence was limited and as yet only a few vertebrae had been found connected together. Nonetheless, from the teeth alone, he said, the ancient beast could be referred to the reptile class and the order of 'Saurians', or lizards.

Buckland outlined the reasons for believing the creature was a lizard. Although he acknowledged that its vertebral column and limb bones were similar to those of mammals, the splendid fossil of the lower jaw betrayed its reptilian origin. Germ teeth showing beside the adult teeth illustrated the replacement cycle characteristic of a reptile and never seen in mammals. Although the beast was found with many marine animals – shells, fish, 'animals of the crab and lobster kind' – there were also amphibian remains such as those of crocodiles and tortoises embedded in the same strata. Consequently, he guessed that the beast 'was probably an amphibious animal'. He envisaged that, unlike the sea lizards, this giant reptile could creep about on land.

Finally, Buckland turned to the question of the lizard's size. Georges Cuvier had compared the ten-inch circumference of the thigh bone to the equivalent bone in modern lizards, and in scaling up their proportions concluded that the ancient beast was forty feet long. Buckland did not disagree with this view, although, he pointed out, 'we cannot safely attribute exactly the same proportions to the recent and extinct species, yet we may with certainty ascribe to it a magnitude very far exceeding that of any living lizard'. In view of the enormous size which this saurian attained, he and the Reverend Conybeare had assigned it the name '*Megalosaurus*'.

Sitting in the audience, eagerly taking in every detail, was Gideon Mantell. He had heard that the Oxford lizard would be announced at this landmark meeting and had travelled to London with fossil bones and vertebrae of the carnivorous lizard from his own collection. Spurred on by Professor Buckland's request to the audience for further information, he rose and, with all eyes on him, explained that he had uncovered fossils of carnivorous teeth in the Weald which matched those from Stonesfield. Even more remarkable, he possessed a giant thigh bone similar to that of Buckland's *Megalosaurus*, which was twice the circumference of the Oxford bone: twenty inches. It would surely be apparent to the audience that if Cuvier's reasoning was adopted, Mantell's specimen might have belonged to a beast almost eighty feet long!

Unfortunately – perhaps in keeping with Mantell's lowly status – the minutes of the Society kept no record of his informal presentation, which is only mentioned in his own diary. There was, however, sufficient momentum and interest in his announcements for the Reverend Buckland to travel two weeks later, 'express from Oxford', to see Mantell's collection for himself. On 6 March 1824 he arrived at Castle Place in Lewes with Charles Lyell, now Vice-President of the Geological Society, and the two men were escorted to the first floor to view Mantell's fine fossil specimens. Buckland could now see the results of all the years of painstaking work: so many fossils so patiently gathered from countless expeditions. He was impressed with the giant thigh bone; the Sussex brute, Buckland thought, must indeed have been of a prodigious size. Buckland and Lyell stayed all afternoon, discussing Mantell's collection well into the evening. The next day, Mantell entered in his diary: 'accompanied Professor Buckland and Lyell to Cuckfield in the afternoon; the rain came.'

As was usual practice, after the meeting at the Society Professor Buckland set about rewriting his talk for publication. According to historian of science Professor Hugh Torrens, there is evidence that when William Buckland prepared his paper on *Megalosaurus* for the records he tried to take advantage of Mantell's discoveries for himself. A stern letter from the Geological Society's publications committee a week later prevented him from going too far. This is the more remarkable since the publications committee was, in effect, warning the new President of the Society not to exploit unfairly the work of another. Mr Warburton of the committee wrote to Buckland on 12 March 1824:

> Whatever you have to say on the subject of the Stonesfield animal, found at Cuckfield, Sussex, must be forwarded at once, since the papers will be required for printing in the next fortnight. I hope that no new plates of the Cuckfield specimens are intended for that paper; it is not a correct practice, and one repeatedly prohibited to other authors to be putting in the last words on the eve of publication; and as President, you are

required to stand by and see fair play to all parties concerned in authorship.

While Mantell was having difficulty getting his ideas past the referees, the publications committee held up the printing of a volume of its *Transactions* in order to incorporate Conybeare's and Buckland's hastily written papers.

Buckland, duly warned of the need for fairness, abandoned his efforts to incorporate illustrations of Mantell's fossils in his scientific paper. When his study was finally published, in more characteristic style, he made generous references to Gideon Mantell, applauding his 'rich and highly valuable' collection and acknowledging that the specimens of *Megalosaurus* in Sussex added to the knowledge of the animal. 'Large as are the proportions of the Oxford individual,' Buckland conceded, 'they fall very short of those which we cannot but deduce from a thigh bone of another of the same species which has been discovered in the ferruginous sandstone of Tilgate Forest.' From the thigh bone found by Mantell, Buckland went on to reason cautiously, the Sussex beast

> must have been twice as great as that to which the similar bone in the Oxford Museum belonged, and if the total length and height of animals were in proportion to the linear dimensions of their extremities, the beast in question would have equalled in height our largest elephants and in length fallen little short of the largest whales. But as the longitudinal growth of animals is not in so high a ratio, after making some deduction, we may calculate the length of this reptile from Cuckfield in Sussex at from sixty to seventy feet.

In fact, these size estimates were wrong. Early geologists were inspired by Cuvier's philosophy, comparing fossils to living animal forms to conjure up the past. This approach had worked well in deciphering the recently extinct mammalian fossils for which Cuvier was now so famous:

the mammoth, *Megalosaurus* and mastodon. Although no one yet realised it, when it came to understanding the ancient reptiles from even further back in time, seeking analogies to living forms was not always the best way forward. It was on this basis that Cuvier had, at first, been doubtful of Mary Anning's *Plesiosaurus* and the idea of a herbivorous reptile. The gentlemen geologists were facing something entirely new and had no frame of reference for dealing with their interpretations.

After the momentous meeting of 24 February, with her *Plesiosaurus* proudly displayed in the Geological Society in the heart of London, Mary Anning's reputation as a fossil hunter seemed assured. Now well known to many collectors, she had become 'the most eminent woman fossilist'. The mineralogist Thomas Allan entered in his journal for June 1824: 'the scientific are entirely indebted to her for the preservation of some of the finest remains of a former world that are known in Europe'. So confident was she of her skills, he added that, 'she is perfectly acquainted with the anatomy of her subjects, and her account of her disputes with Buckland, whose anatomical science she holds in great contempt, was quite amusing'.

Mary Anning's friend Anna Maria Pinney described this self-confidence differently: 'She has been noticed by all the cleverest men in England who have her to stay at their houses, correspond with her on Geology &c. This has completely turned her head and she has the proudest and most unyielding spirit I have ever met with. Much "learning has made her mad". She glories in being afraid of no one and in saying everything she pleases. She would offend all the world, were she not considered a privileged person.' This, however, does not altogether tally with reports of another side to her character. She was respected locally because 'she were kind to the poor . . . whenever she found a little cag [keg] on the shore, she would cover it up, and not let the preventative men see it, but would tell some poor person of it'.

In contrast to the recognition that Mary Anning received, Gideon Mantell was still frustrated by his lack of progress. His independent discovery of the bones of a giant carnivorous lizard had, almost certainly, stimulated William Buckland's scientific paper. Yet as the

Reverend Buckland was the first to provide a name and a full scientific account, he naturally received the credit. Mantell's paper identifying the Tilgate beds of the Weald as Secondary rock was not yet accepted or published by the Geological Society, and consequently his interpretation of the worn herbivorous teeth as those of an ancient lizard was still in dispute.

'As if to add to the difficulty of solving the enigma,' Mantell wrote, a 'tubercle' or horn was uncovered in the same strata, 'equal in size, and not very different in form, to the lesser horn of the Rhinoceros. It is externally of a dark brown colour; and while some parts of its surface are smooth, others are rugous and furrowed as if by the passage of blood vessels . . . The horn . . . was declared by competent authorities to be the lesser horn of a rhinoceros.' Both the worn herbivorous teeth and the horn appeared to confirm the conclusions of the experts, that Mantell had uncovered nothing more than a rhinoceros.

Although Professor Buckland did not accept that Mantell had found an ancient herbivorous lizard, by the spring of 1824 he probably did concede that the Tilgate beds in the Weald were Secondary rock. Indeed, Buckland could hardly disagree when presented with so many fossil bones from Tilgate that matched those of the *Megalosaurus* from the ancient Secondary rock at Stonesfield in Oxfordshire. However, since Georges Cuvier in Paris had declared that the unknown herbivorous teeth belonged to a rhinoceros, Buckland became trapped in a circular argument. Not liking to take a view that would appear in disagreement with Cuvier, he now maintained that the disembodied herbivorous teeth had come from a younger deposit near the surface and had fallen down a fissure in the rocks into the older strata. These unknown herbivorous teeth, he maintained, should still be assigned to some recent mammalian creature or fish. Unfortunately for Mantell, he could not disprove Buckland's conclusion. Many of the teeth had been found loose, some-times even scattered on the roadside where quarrymen had dumped stones, so it was difficult to persuade Buckland that they had ever been properly embedded in the ancient Secondary rock.

*   *   *

By now, Gideon Mantell's youthful optimism that he could straddle two careers and make progress in science by sheer hard work had been somewhat shaken. The relentless demands of his medical practice in Lewes continued unabated, even with additional help from his young brother Joshua. 'Exceedingly engaged,' he wrote in his journal on one occasion, 'my brother and myself visited ninety patients'. The nights were still punctuated by urgent cases of midwifery, the daytime visits by distressing emergencies: 'Requested to attend an accident at Malling Mill; on my arrival I found a poor boy with his hand dreadfully lacerated; thought it necessary to remove the forefinger and thumb at the wrist joint,' Mantell wrote in his diary. Another time he attended 'a severe case of fractured skull at Barcombe; was obliged to perform the operation of trephining: the man not likely to recover'. He had success with many operations, nonetheless – fractures of the collarbone, of the legs, even delicate surgery such as removing cataracts. His efforts for the poor were always appreciated: 'received a vote of thanks for the skill and attention for the poor of St John's'.

Occasional visits from the gentry to view his geological collection kept Mantell's hopes of patronage alive: 'The Earl of Chichester called and presented me with a fine antique gold ring, which had been found at Bormer; his Lordship was very affable and polite and presented Mrs Mantell with some Irish Gold.' His collection of curiosities was also seen by others of high rank, local gentry such as Lady Gage and Sir George Shiffner, and visitors from further afield including Mary Shelley, the wife of the famous poet and author of *Frankenstein*. Each time-consuming visit was politely and eagerly attended by a hopeful Mr or Mrs Mantell.

Despite his ambivalent feelings about the progress of his geological research, Mantell was quite unable to give it up. There were bones scattered in every available space in the drawing-room: they lay in tray upon tray, piled in heaps on the floor and on shelves that reached the ceiling, the light hitting them variously at different times of day and emphasising their inhuman shapes. The entire scene was shrouded in an unwholesome musty aura of ancient bones. And there in the middle of it all, totally absorbed in his studies, was the eminently sensible Dr Mantell.

Having obtained detailed drawings and engravings of the new sea lizard, the *Plesiosaurus*, and the giant *Megalosaurus* from the papers published by the Geological Society, he now set about comparing all the bones in his collection and more accurately reassigning some of them to the newly identified animals. Mantell had long known that some of the fossil teeth, by their very distinctive conical shape, resembled an ancient species of crocodile. He had assigned the enormous thigh bone, the blade-like carnivorous teeth and a few vertebrae to the Oxford carnivore, the *Megalosaurus*. And there were other vertebrae and teeth that matched those of the *Plesiosaurus*. But in this ever-expanding jigsaw that had taken over his drawing-room like some slow fungal growth, there were pieces remaining, mostly giant vertebrae and teeth, that did not fit anywhere. Could these fossils belong with the herbivorous teeth?

'By stimulating the diligent search of the workmen by suitable rewards,' Mantell wrote, he obtained many more of the unknown herbivorous teeth. With each new specimen, he was struck by a feature that had not been seen before. The young teeth had marked *serrations*, all the way round the crown, that bore no resemblance to the teeth of a wolf-fish or a rhinoceros. Some of the young teeth were so perfect that they showed the 'serrated edges, longitudinal ridges and the entire form of the unused crown'. By the spring of 1824, Mantell was becoming certain that the Reverend Buckland was wrong in arguing that the herbivorous teeth came from a recent mammal or a fish.

Although he had many disappointing specimens – a fang broken or a cavity filled with sandstone – gradually the pattern of dentition of the ancient beast began to emerge. He suspected that as the crown of each young tooth had been ground down with wear, the characteristic serration on the ridges of the enamel had been obliterated, leaving eventually a mere worn-down stump in a mature tooth. It occurred to him that some of the fangs had a cavity or depression at the base, perhaps caused by the pressure of a secondary tooth growing through below. In one specimen the cavity at the base of the fang, for the reception of a new tooth, was remarkably distinct. Even without a fossil jaw, if he could prove that there was a replacement cycle, with new teeth growing through to replace the worn ones, then, he reasoned, the creature was surely reptilian? Over the months, he accumulated an entire series of teeth showing 'every graduation of form from the perfect tooth in the young animal to the last stage, that of a mere bony stump worn away by mastication'. Georges Cuvier had seen the mature worn teeth, but what would he make of the younger teeth with their distinct serrations?

In the spring of 1824, Gideon Mantell felt optimistic enough to try his luck with Cuvier again; he made drawings of the entire series of teeth and sent them to Paris. He submitted some fossils and drawings of the *Megalosaurus* from Sussex for Cuvier's new book, and explained that the size of the larger specimens made them difficult to send. He then added: 'I have enclosed a few specimens of the teeth which Mr Lyell submitted to your notice earlier . . . These teeth have not been discovered in any other part of England and our comparative anatomists have not been able to afford me any information respecting them. I have therefore been most anxious to place them in your hands confident that they will be fully investigated.'

Correspondence shows that Cuvier was initially puzzled by this series of teeth. Buckland wrote to him on 2 June 1824, stating his opinion that they were from 'the anterior part of the jaw of some fish as in the Ballistes and Tetradon'. Eventually, on the 20th, Mantell received a reply from Paris. As he opened the envelope he could see at once the distinctive and elegant hand of the Baron. The letter was in French, so he could not interpret it at a glance: 'Ces dents me sont certainement inconnues; elles ne sont point d'un animal carnassier, et cependant je crois qu'elles appartiennent . . . '

Gradually, it began to dawn on Gideon Mantell that it was good news. 'I hasten to express my gratitude [for the fossil specimens] and to offer a few ideas inspired by the curious teeth which are part of your package,' Cuvier wrote. 'These teeth are certainly unknown to me. They are not from a carnivore; nevertheless I believe . . . they belong to the order of reptiles. From their exterior one could also take them for fish teeth, similar to tetradons or diadons, but their external structure is very different. Might we not have here a new animal, an herbivorous reptile?'

Even though he had only a series of teeth to go on, the Baron, at last, gave him the encouragement he needed.

'Some of the great bones that you possess should belong to this animal which, at present, is unique of its kind. Time will confirm or crode this idea, since some day a portion of the skeleton may be found with parts of the jaws carrying the teeth. It is this last object, above all, which must be searched for with the greatest perseverance. If you could obtain some of these teeth still adhering to a small portion of the jaw I believe that the problem would be solved.' He concluded by telling Mantell that he would be doing him 'a very great service' if he informed him of any further developments.

Mantell's vision of a giant herbivorous lizard roaming the ancient world, an idea held for so long against general criticism, had proved to be right all along. For him this was the turning-point; with Cuvier's support, his views could no longer be discarded so lightly by the English 'experts' in London. 'Nothing could be more gratifying to me, than your confirming me in the opinion that the curious teeth of Tilgate Forest belong to an unknown animal,' he replied jubilantly to Cuvier on 9 July, 'and are not those of the Diodon as Professor Buckland and others of my friends would insist upon.'

To find out what kind of reptile it could be, early in September Mantell travelled to London with some of the teeth specimens, to visit the Hunterian Museum at the Royal College of Surgeons. His aim was to find out if any living reptile had similar teeth. This could shed light on what kind of jaw the ancient reptile possessed and what type of beast within the reptile class it might be. The museum was open only twice a week, for a few months of the year. William Clift the Curator was still struggling with a backlog of anatomical specimens to identify. Nonetheless, he welcomed Mantell and made time to help with his search. Together they ransacked drawer after drawer of fossils, looking for others with features comparable to the unknown reptile teeth. They could find nothing with similar strange serrations. But their scholarly discussion attracted the attention of others including the assistant curator, Samuel Stutchbury.

Stutchbury had worked in the Bristol Museum, where curios and oddities from the West Indies were occasionally brought back by slave ships returning with their cargo of sugar and cotton. As a result, he had some knowledge of living tropical animals. It occurred to him that Mantell's fossils were not unlike a giant version of a modern lizard: the Iguana. By sheer good fortune, he had just finished preparing in alcohol a specimen from Barbados at the Royal College, and offered to show it to them.

When the two sets of teeth, from the fossil and the living animal, were directly compared the similarities were striking. In both animals, Gideon Mantell wrote, 'the edges are strongly serrated, the outer surface is ridged, the inner smooth and convex; the secondary teeth appear to have been formed in a hollow in the base of the primary ones, which they expelled as they increased in size . . . The teeth appear to have been hollow in the young animals and to have become solid in the adult.' It was an intriguing set of resemblances.

The rather drab hall of the Hunterian Museum, full of long-dead specimens gathering dust, was in stark contrast to the lively interest of the three men who, busy with their magnifying glasses, were totally absorbed in the intricate details of their science. They did find a few subtle differences between the iguana and the fossil teeth: the raised indentations and markings didn't follow quite the same pattern. Nonetheless, it was soon clear that there was only one major difference: size. The fossil teeth were twenty times bigger than the modern iguana teeth. For Mantell, this raised another exciting possibility. The little modern lizard in front of them stood just over three feet high. Scaling up from the size of the teeth, his ancient animal might be twenty times larger: a preposterous size, in fact – perhaps sixty or seventy feet long – a lizard that would have stretched along the Lewes High Street almost from the Castle to St Mary's Lane.

Shortly after this meeting, in September 1824, Mantell began to prepare a scientific paper on his new animal. He wrote to William Clift:

'since I had the pleasure of seeing you in Town, I have endeavoured to obtain a specimen of the Iguana Tuberculata, that I might introduce a sketch of the jaw to illustrate the history of my herbivorous reptile . . . may I intrude on your indulgence and beg permission to make a sketch of the lower jaw of the Iguana in the museum?' Clift, who was a skilled draughtsman and only too keen to help, obligingly made beautiful drawings of the jaw of the modern iguana, magnified four times to show the serrated edges, for Mantell's paper.

There was just one outstanding question: what to call the newly identified reptile. Mantell wrote to Cuvier on 13 November, to inform him of the similarities to the modern iguana: 'so striking is this analogy, that I have ventured to propose the name of *"Iguana-saurus"* for the fossil animal,' he said. However, he also received advice from the Reverend Conybeare. 'Your discovery of the analogy between the Iguana and the fossil teeth is very interesting,' Conybeare wrote, 'but the name you propose "Iguana saurus" will hardly do, because it is equally applicable to the recent Iguana. *"Iguanoides"*, like an Iguana, or *"Iguanodon"*, having the tooth of the Iguana would be better.' By December 1824, Mantell seems to have resolved the matter and settled on Conybeare's advice: the '*Iguanodon*'.

Mantell had another stroke of good fortune in the autumn of 1824, when Georges Cuvier finally published his new edition of *Recherches sur les Ossemens Fossiles*. In this volume he was prepared to admit he had made an error in identifying the *Iguanodon* tooth. He explained that at first sight he had mistaken the worn-down tooth for an incisor from a rhinoceros, and he pointed out the great debt he owed Mantell: 'it is only since Monsieur Mantell sent me an entire series [of teeth], more or less used, that I am quite convinced of my mistake'.

With such public acknowledgement from Cuvier, Mantell was hastily admitted into the elite circles of the London societies. William Buckland wrote, urging him to attend the next meeting of the Geological Society and inviting him to dine beforehand with a few others in a tavern in St James Street. His fame even reached the Continent: on 28 November 1824, he entered in his journal: 'during this last week, have had

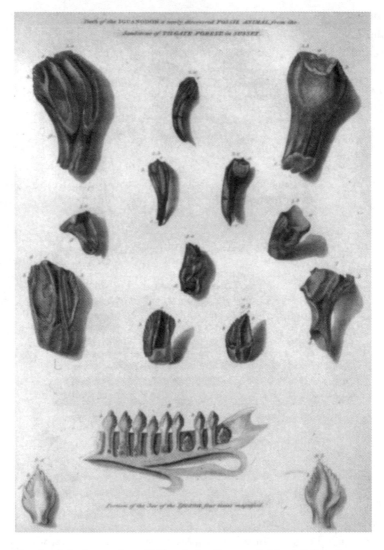

Teeth of the *Iguanodon*. Although the first teeth Mantell found were
very worn, he eventually collected a series including young teeth
with marked serrations.

numerous applications from different persons respecting the new animal whose teeth I have discovered in the sandstone and which I have named the Iguanodon. Have prepared parcels for Cuvier, Brongniart, Schlotheim, and Baron Humboldt and two packets of fossils for Dr Brown of Glasgow.'

Cuvier's publication also changed William Buckland's fate, and in an unexpected direction. According to an account in the family archives, 'William Buckland was travelling in Dorsetshire, reading a new and weighty book of Cuvier's which he had just received from the publisher; a lady was also in the coach and was reading this identical book, which Cuvier had also sent to her.' In spite of there being no one there to introduce them, the magnetism of the two books proved irresistible. Intrigued, they soon found themselves deep in conversation, 'the drift of which was so peculiar, that Buckland at last exclaimed, "You must be Miss Morland, to whom I am about to deliver a letter of introduction."' She was indeed Mary Morland, an established fossil geologist in her own right. She shared Buckland's love of natural history and, it seems, felt quite at home in his Christ Church menagerie at Oxford, complete with Tiglath and the other exotic pets. Buckland, now a confirmed bachelor of well over forty, speedily married Miss Morland. The wedding tour was a geological excursion of nearly a year around Europe.

Buckland had made a most fortunate choice: 'Not only was she pious, amiable, and an excellent helpmate to my father,' commented her son Frank, 'but being naturally endowed with great mental powers, habits of perseverance and order, tempered by excellent judgment, she materially assisted her husband in his literary labours and gave to them a polish which added not a little to their merit.' To add to her charms, she was 'devoted' to her husband's pursuits; 'clever in mending fossils from a mass of broken and comminuted fragments . . . she labelled them in a particularly neat way and there is hardly a fossil or bone in the Oxford Museum which has not her handwriting upon it'. On top of all these sterling qualities, she also 'occupied herself in schemes of charity . . . and did not neglect the education of her children'.

Meanwhile, Gideon Mantell could at last perceive a focus for his ambition. Perhaps because the members of the Geological Society had opposed him for so long, he set his sights on presenting a scientific paper to the most important scientific body of all: the Royal Society. Sir Isaac Newton had been its first President, and its members were renowned as gentlemen of 'exceptional ability . . . dedicated to the untiring pursuit of knowledge'. Membership of this prestigious society was becoming a way of improving social position, gaining status and, best of all, of meeting wealthy nobility with an interest in the sciences who might possibly serve as patrons. The passport for someone from a humble background to this world of opportunity was to be made a 'Fellow' of the Royal Society, which was granted only if the applicant was nomin- ated by a member and had distinguished himself greatly in some aspect of science.

Gideon Mantell approached his friend Davies Gilbert, now Vice- President of the Royal Society, to help advance his case. He wrote to Gilbert on 1 January 1825 outlining his discovery of the *Iguanodon*, and soon heard that this letter would be read out as a scientific paper before the Society, on 10 February. The meeting was to be held in the grand assembly hall of Somerset House in the Strand. Mantell himself was unable to attend.

The large and airy meeting-room, graced with the gilded portraits of former dignitaries, echoed with the gossip of the gathering assembly. As the President, Sir Humphry Davy, famed for the creation of the miner's safety lamp, took the chair, silence descended. The Secretary opened the meeting with the usual remarks: 'The following strangers have leave to be present . . .' – whereupon, those who were not Fellows of this august institution were formally introduced. Then the minutes of the last meeting were read. 'The following presents were received and thanks ordered for them,' continued the Secretary. Memoirs, papers, journals and illustrations received by the Society were duly noted. 'Next, we have a Certificate to present in favour of Reverend Robert Morrison of Canton in China, to be suspended in the meeting room . . .'

At last, the President introduced Davies Gilbert, Esq., who then solemnly rose and began to read Mantell's paper, 'Notice on the

Iguanodon, a Newly discovered Fossil Reptile from the sandstone of Tilgate Forest':

> 'Sir,
> I avail myself of your obliging offer to lay before the Royal Society a notice of the discovery of the teeth and bones of a fossil herbivorous reptile in the sandstone of Tilgate forest; in the hope that, imperfect as are the materials at present collected, they will be found to possess sufficient interest to excite further and more successful investigation . . .'

Mantell went on to outline in his letter the lengthy detective work that he had undertaken to identify the animal. 'But although my communications were acknowledged with that candour and liberality which constantly characterises the intercourse of scientific men, yet no light was shed on the subject, except by the illustrious Baron Cuvier.' Gilbert read out an extract of Mantell's correspondence in French: 'N'aurions-nous pas ici un animal nouveau, un reptile herbivore?' Mantell also acknowledged William Clift, 'to whose kindness and liberality I hold myself particularly indebted', for his help in comparing the fossil teeth. He explained that, because of the freshwater fossils with which the remains were associated, the beast was 'not of marine origin'. With considerable satisfaction, he went on to name the animal: 'the term Iguanodon, derived from the form of the teeth . . . will not, it is presumed, be deemed objectionable.'

However, since they only had teeth to go by, he said, there was a critical dilemma.

> 'Until some connected portion of the skeleton shall be discovered it is impossible to distinguish the bones of the one from those of the other. Since, however, the teeth of the Iguanodon are not known to occur in the Stonesfield [Oxford] slate perhaps such of the bones from Tilgate Forest as resemble those described by Professor Buckland may be attributed to the

Megalosaurus, while others not less gigantic may be assigned to the Iguanodon. That the Iguanodon equalled, if not exceeded the former in magnitude seems highly probable, for if the recent and the fossil animal bore the same relative proportions, the tooth must have belonged to an individual upwards of sixty feet long!

I have the honor to be, Sir

your most obedient Servant

Gideon Mantell'

For Mantell, it was a significant advance. No one had queried the opinion of an amateur to name a new reptilian species. No one had challenged his view of the extraordinary size of the beast. Without any institutional backing or the reassuring support of a close circle of colleagues, he had successfully convinced some of the most learned men in the country of his discovery. In a matter of weeks, he was not only elected a member of the Council of the Geological Society, but he also received a letter announcing his election as a member of the Philomathic Society of Paris.

But, eclipsing everything, the honour that meant the most to Gideon Mantell finally came his way. He was informed that several distinguished scientists, including William Clift and William Buckland, had nominated him to be a Fellow of the Royal Society; he was recommended as a 'Gentleman well skilled in science and particularly in geology and well known for his remarkable paper on the Iguanodon'. Then on 22 December 1825 Mantell was invited to attend an evening ceremony at Somerset House, to be formally admitted to the Society. In the grand assembly rooms he met Mr John Herschel, son of the famous astronomer Sir William and a distinguished Cambridge mathematician, along with his friend Mr Charles Babbage who later designed the first calculating machine. Babbage and Herschel introduced Mantell to the other Fellows, and then he was requested to sign the 'obligation' in the Society's charter. From the yellowed parchment he read: 'We who have here unto subscribed, do hereby promise each for himself that we will endeavour to promote the good of the Royal Society of London for improving Natural Knowledge and to pursue the ends for which the same was founded.'

He was then escorted with the other Fellows to the grand meeting-room, where Sir Everard Home, 'the London Baronet' and friend of the Prince Regent, was in the chair presiding over events. The Secretary rose to begin the proceedings in the centuries-old tradition: 'The following strangers have leave to be present . . . ' Various non-members were then introduced, the minutes of the last meeting were read, presents gratefully acknowledged and important messages delivered. Finally, Sir Everard rose and turned to Mantell. 'Gideon Mantell Esq having . . . signed the obligation in the Charters book is admitted to the Royal Society.'

There were smiles and a murmur of approval. The gentlemen next to him shook his hand, and he was invited to take tea in the library with the other Fellows, Sir Anthony Carlisle, senior Council member of the Royal College of Surgeons, Charles Babbage, John Herschel and others. The party also included titled aristocracy, who would surely provide Mantell with patronage. As he sank into the soft leather chair, surrounded by walls lined with faded bound volumes and sipping tea from fine bone china, he could muse with some satisfaction on the events that had brought him to this point. Here he was, the shoemaker's son, now established as a scientist and a gentleman, at the heart of it all, in the most prestigious society in England. He could enjoy at last the glory of being proved right over other, more eminent men – arguably, even, the famed Georges Cuvier.

All his hunches had been right, his advisers had been wrong. He, Gideon Mantell, had uncovered some of the best specimens of the only two great land reptiles known to Man. He had identified an entire herbivorous species, *Iguanodon*, from an incredibly small clue, a tooth. In addition, it was his fossils that had stimulated the first paper on *Megalosaurus*, the great carnivorous reptile, and shed light on its size. With masterly insight, he had finally established his supremacy in this field of study; and all the while working as a country doctor.

At last, he felt he had the blue ribbon of science. He could glimpse the promised land; the life of learned men exploring new intellectual territories in all the emerging fields of science. As John Herschel observed, geology was fast becoming comparable to astronomy in the awesome 'magnitude and sublimity' of the ideas raised: the vast, strange history of the globe was opening up before their eyes. Mantell wrote in his diary: 'it was with no small degree of pleasure that I placed my name in the Charter book which contained that of Sir Isaac Newton and so many eminent characters'. He, too, wanted to illuminate ignorance and reveal the glorious wonder of Creation. He longed to understand 'the prolific productions of Nature in chronological order' and 'the history of thousands of centuries that preceded human existence'. If Newton was the greatest luminary, Mantell hoped that perhaps his name, too, might now light up the pages of history.

PART TWO

6

The Young Contender

From millions take thy choice,
In all that lives a guide to God is given;
Ever thou hear'st some guardian angel's voice,
When nature speaks of heaven!

Cited in *Thoughts on a Pebble*
by Gideon Mantell, 1849

In little more than a decade, William Buckland's humble study of 'undergroundology' had blossomed into the 'Queen of Sciences'. Geologists had mapped the order of succession of Secondary and Tertiary rocks in England and had begun to recognise the different periods that formed the 'pages of history' of the ancient globe. Remarkable fossil discoveries by dedicated collectors had revealed that strange beasts, quite unlike any living animals, once inhabited the world. There were Mary Anning's monstrous sea lizards, the ichthyosaurs and plesiosaurs that swam in primeval oceans. Equally bizarre were William Buckland's forty-foot carnivorous *Megalosaurus* and Gideon Mantell's gargantuan herbivorous *Iguanodon*, which was variously estimated at between fifty and a hundred feet long.

Yet by the mid-1820s, the emphasis in geology was gradually shifting. Rather than an intimate knowledge of the rock, to understand the fossil animals of these 'former worlds' a new set of skills was becoming important. In 1825, the year that the thirty-five-year-old Gideon Mantell finally established himself as a gentleman scientist at the Royal Society, a young man arrived in London who was to have a significant role in interpreting fossil reptiles, although he was no geologist and had never set foot in a quarry. His name was Richard Owen, and he was an anatomist.

Even as a young man, Owen made a striking impression. He was very tall and, according to Thomas Carlyle the philosopher and essayist, possessed 'great glittering eyes'. An oil painting of Owen at the time suggests that his eyes were indeed a noticeable feature: large and dark, with a penetrating intensity. His straight hair is neatly swept across a broad expanse of forehead, framing his face in the fashionable sideboards of the day. A pressed, upturned white collar partially conceals a prominent chin, and the sombre colours of his suit form a marked contrast to his pale complexion.

Richard Owen's enthusiasm for study was not a feature of his childhood. He came from the northern town of Lancaster, where his family owned a five-storey house on the edge of Dalton Square. His father had capitalised on the new development of woollen mills and canals at the turn of the nineteenth century to expand his draper's business. He had amassed a considerable fortune in trade with the West Indies, although he had later lost some of it when Napoleon reneged on all France's debts to the British. The prosperous Owens had been keen to obtain a good education for their two sons. Undeterred by the fact that Richard saw no point in his lessons, his ambitious mother, Catherine, enrolled him at the Lancaster Grammar School at the tender age of six.

The family archives show that the ebullient young Owen's most enduring memories of his schooldays were of the 'Black Mondays'. This was the day when the misdemeanours of the previous week were publicly announced and suitably hideous punishments, designed to form character, were duly inflicted. Richard was not beyond getting embroiled in fights with the brightest boy in the school, William Whewell, and occasionally coming home with a black eye. Whewell was later to become a celebrated Master of Trinity College, Cambridge. By contrast, the tutor held out no hope for Richard Owen: he was 'lazy and impudent' and would 'come to a bad end'.

In 1820, after ten years at school, the sixteen-year-old Owen was unwilling to settle to any particular trade. That same year he signed up as an apprentice to a local surgeon, Mr Leonard Dickson, whose medical practice extended to the prisoners in the county gaol sick-rooms. Armed with little more than a few leeches, the doctor and his young apprentices did daily battle with consumption, cholera, smallpox and many other killer diseases. Overcrowding in the prison was normal, and the dark, lice-scabbed walls encased such misery that it was hard to see how anyone could survive. It was in this gruesome setting that Owen received his introduction to the science of anatomy, through post-mortem work on unfortunate inmates who had died.

The first autopsy that Owen attended was such a shock that he almost gave up his apprenticeship there and then. Merely entering the gaol was frightening enough: the forbidding fortress was a towering edifice on a hill, built with turrets and ramparts. Owen and his master entered through a great portcullised gateway, guarded by a turnkey. Obediently, he followed the surgeon across the yard and up an endless spiral of stone-flagged steps to a room in the old tower, which was used as the prison washroom. With a sense of disbelief, he saw the contours of human bodies lying on stone slabs under white sheets.

As one sheet was flung back, Owen was alarmed to see 'the pale, cold collapsed features of the deceased, the half opened eyes . . . the glassy staring eyeballs'. It was the body of a young man whom he had tried to save only a few days previously. He felt 'over-awed by the power of the human corpse', and revolted and distressed by the surgeon's clinical dissection, which seemed a terrible 'desecration of the sanctity of the dead'. The youthful and naïve Owen, who had no doubt that even muttering the Lord's prayer backwards as a spell could 'raise the Devil' and who held due reverence for a score of similar schoolboy myths, found his enthusiasm for science 'damped considerably'.

By chance, at nine o'clock that same November evening, Owen was summoned to the gaol again to deal with a few cases of fever. As he made his way across town, a storm was rising. It was dark by the time he reached the portcullis and raised the heavy iron knocker. Once more, lantern in hand, he crossed the cobbled yard to the turret that led to the tower. As he turned the heavy key, he was knocked aside by a great gust of wind. The lantern swung open and his light was extinguished. Trapped on the spiral stairway as the door to the tower closed behind him, he was plunged into pitch-black.

'The loneliness of my position first then struck coldly upon me,' he recalled. Each step brought him closer to the cold chamber where the dissected corpses lay stretched out on the stone slabs under their white sheets. As he passed the chamber, a sound made him raise his head. 'My alarm grew into a creeping, freezing horror, as I, staring intently upwards, made out by degrees the pale, collapsed features and those half opened glassy eyes that had haunted me through the day and now looked coldly down and met my own.' The thin figure now clasped the central pillar of the staircase. The young Owen rushed to make a descent, but 'I had hardly made one turn down past the closed door of the dead-chamber when a second figure in white appeared below me as if to intercept my passage . . . and surely it bore the features of the other corpse! I grasped the pillar for support and gazed upon the spectre in speechless terror.'

While anyone else might have fled from the dark gaol as fast as he could, Owen showed a presence of mind that tells us something about him. As he began to make his escape, he suddenly became aware that his feet were dragging some strange object. At the first thin shaft of light, he could see he had a white sheet entangled around his feet. This simple 'evidence of materiality' seemed to pull him to his senses. Summoning all his courage, he went back up the stairs and saw in the moonlight falling through the arrow-slit window that a nail in the wall had been used to hang the prison sheets to dry. In his terror, he had somehow mistaken the hanging sheet for a ghost. He put the sheet back and continued up the stairs to investigate carefully the source of the first ghost. This mystery, too, he could explain by the way the light fell through the higher window.

Having solved the problem he made his exit, nodding politely to the turnkey and summoning an air of authority and dignity as best he could. However, once out of sight, he collapsed with the shock. When he recovered, he found himself vowing all the way home, 'never, never, again to desecrate the Christian corpse and to quit a profession that could only be learned by such practices so repugnant to the best feelings of one's nature'.

However, within six weeks Owen's fascination with anatomy had become so compelling that all his other fears and scruples were brushed to one side. Encouraged by fellow pupils and excited by some articles he had read, he had begun to form a small anatomical collection, including the skulls of dogs and cats and the skeletons of mice. But this tame collection of domestic animals was soon insufficient to satisfy his appetite for knowledge.

By chance, a black patient had died in the gaol hospital and Owen assisted at the post mortem. Inspired by an article he had read on 'The Varieties of the Human Race', he slipped some silver to the old turnkey. 'I told him I should have to call again that evening to look a little further into the matter before the coffin was finally screwed down.' It was snowing that night, when he returned to the gaol. He made his way up the same spiral stairway that had so terrified him just a few weeks previously, entered the corpse room and took the head of the dead man. Carefully concealing the head in a brown paper bag under his cloak, he went back down, past the turnkey. His thoughts, he said later, were only on craniological speculations of 'facial angles', 'prognathic jaws' and the 'peculiar whiteness of osseous tissue'.

But his thoughts were not on such lofty matters for long. As he hurried down the hill, he slipped on the ice and lost his balance. The black head was catapulted out of the bag and went bounding off down the slippery hill, pursued by Owen in his great, flapping dark cloak and leaving splashes of red on the white pavement slabs. It bounced against the door of a cottage, which flew open, and he heard unearthly shrieks from inside. Owen rushed inside, 'saw the whisk of a garment of a female' vanishing through the door, 'and the ghastly head at my feet with its white protruding eyeballs'. He grabbed it and ran home.

The next day the whole town was talking of the phantom, which was widely rumoured to be the ghost of a Captain Tasker and his Negro

The young Richard Owen.

slave, perhaps even the Devil himself. For any doubters, a drop of blood, now dry and dark by the door to the cottage, provided proof of their nocturnal visit. Only Richard Owen knew the truth about the terrifying apparition that haunted the town, and for years he did not 'disburden' himself of what had happened.

In a matter of weeks, it seems, the sixteen-year-old Owen, spurred on merely by some articles in a cyclopedia, was transformed from an apprentice ruled by schoolboy superstitions into a young man who would stop at practically nothing to improve his skill in dissection and make headway in his career. Thefts of body parts, bribery of officials, even ghostly fears, were no longer an obstacle. The fledgling subject of anatomy held the young Owen in thrall.

In 1824, shortly after the Reverends Buckland and Conybeare had announced the existence of the *Megalosaurus* and *Plesiosaurus* to the Geological Society, money was found to send Richard Owen to Edinburgh University to study medicine. Scotland's capital had become during the second half of the eighteenth century one of Europe's most cosmopolitan and fashionable centres. As Owen's carriage turned into Princes Street, he could glimpse down the side-roads the elegant, recently built squares of the New Town and the classical homes of the prosperous merchants. Across the great ravine that divided the town, beyond the steep alley-ways and precipitous steps leading up to the stately High Street, stood the castle, perched on a massive rock that rose high above the New Town, the law courts with their bewigged judges hurrying down Canongate, and a cluster of schools and Gothic buildings around the university seething with student life. It was here that Owen settled into lodgings in Nicholson Street and began to register for courses.

Since the Scottish Enlightenment when the ideas of philosophers such as David Hume, William Robertson and Adam Smith had influenced intellectuals across Europe, Edinburgh had become renowned as a centre of learning. Although the 'golden age' of Edinburgh had ended by 1800, because of strong links with European universities the medical training at Edinburgh was still broader than any in England. In the 1820s, Edinburgh offered more courses and greater awareness of continental thinking than anywhere else in Britain. Many medical students visited Paris – not least because of the availability of cheap cadavers at French hospitals – and came back with news of all the most radical European ideas.

Despite his deep interest in anatomy, Owen spurned the lectures of the resident professor of anatomy, Alexander Monro the Third, who seemed a curious throwback to the eighteenth century. Monro had inherited both his eminent chair and his lecture notes from his father and grandfather. By all accounts, he cut a shabby, dirty figure; it was not uncommon for him to enter the lecture hall spattered in blood from dissections. His teaching was woefully out of date and there were times when his sessions turned into student protests. It was Monro's lectures that were to stir such a powerful aversion to human anatomy in the young Charles Darwin the following year.

Richard Owen was much more excited by the extramural anatomy lectures of Dr John Barclay, who had been teaching anatomy in Surgeon's Square in Edinburgh for nearly thirty years. Barclay's lectures were recognised by the Royal College of Surgeons in Edinburgh and described as 'of a very superior order to that of the third Monro'. Enveloped in the pungent smell of preservative in the claustrophobic rooms of Barclay's anatomy school, Richard Owen soon realised that anatomy was not just the practical tool of the surgeon for understanding the workings of the body and the causes of death. It was a powerful agent for tackling fundamental issues in science: the origin and extinction of species, the process of Creation itself. And nowhere were the ideas of 'philosophical anatomy', as it became known, more hotly debated than at the Muséum National d'Histoire Naturelle in Paris.

Since propounding his early evolutionary ideas at the turn of the century, Jean-Baptiste Lamarck, Professor of Invertebrates at the museum, had been eclipsed by his politically astute younger colleague, Georges Cuvier, who rose to much greater fame. Cuvier had continued to oppose Lamarck's view that living forms could be modifications of ancient creatures. Undeterred, between 1815 and 1822 Lamarck embarked on a definitive restatement of his theories in his seven-volume study on invertebrate animals, *Histoire Naturelle des Animaux sans Vertèbres*. In his work, Lamarck proposed that Nature has a tendency towards the increasing complexity of animal forms. Changing environments prompted new responses from animals, which could become habitual. In turn, animal organs were modified by frequent use, allowing for the progression of animal species.

For Lamarck, who was also battling against failing eyesight, the effort of producing these volumes was considerable. By 1818, he was completely blind and could only finish the project by dictating to his daughter. Although this work is now described by academics as one of the 'glories of French science', it made little impact on mainstream science at the time. When Lamarck died a few years later, his family were so poor that there was no money for his funeral and they were

Jean-Baptiste Lamarck.

obliged to petition the Académie des Sciences to pay for his burial.

Jean-Baptiste Lamarck, however unfashionable, was not without disciples. The most outstanding among them was Étienne Geoffroy Saint-Hilaire, a man nearly thirty years his junior. Geoffroy had been appointed to the prominent position of Professor of Vertebrates at the

Muséum National d'Histoire Naturelle in Paris in 1793, when he was just twenty-one. He joined Napoleon's Egyptian campaign at the turn of the century and soon made his name with a series of studies on the crocodiles of the Nile. His brilliant lectures and his contribution to zoology earned him the admiration of the writer Honoré de Balzac, who called him 'the Grand Marshal of the Grand Army of Philosophers'.

During the early nineteenth century, Geoffroy pursued and developed Lamarck's ideas with a vigour that inevitably pitched him firmly against Georges Cuvier. For Cuvier, God had created species for a particular role in Nature. Since this role was fixed from the beginning, there was no need for species to vary; evolution made no sense of God's purpose. In Cuvier's mind, the idea that Nature was autonomous and could almost randomly generate higher forms of existence – without God – was unthinkable. But Geoffroy would not be silenced by these assertions. He was particularly interested in how species might be modified by their environment, and was searching for proof of Lamarck's view that species are transformed over time from simple animals into higher forms of life. In 1822, Geoffroy published *Anatomical Philosophy*, in which he highlighted puzzles in anatomy that lent weight to the evolutionary ideas of Lamarck.

A powerful line of argument came from the study of 'homologies'. The wing of a bat, the fin of a fish and the hand of Man, for instance, are homologous in that they occupy equivalent parts in the different animals. While Cuvier maintained that a fish's fin bore no relation to a mammal's paw, each having been designed separately by God, for Geoffroy homologous organs highlighted important relationships: the bat's wing, the animal's paw and a man's hand had the *same* bones. Such observations led him to believe that all vertebrate animals were constructed to a single underlying plan. This 'unity of plan', by showing a close relationship or kinship between different vertebrate animals, added plausibility to his view that the higher creatures might have evolved from lower ones.

To demonstrate how mammals, for example, might have evolved from fish, he studied developing embryos. It was already known from eighteenth-century studies in Germany that vertebrates, during their early life as an embryo, pass through several different stages that resemble more primitive forms of life. This was not easy for Creationists to explain. If God made each and every creature, why should the human embryo, in its early stages, possess the initial forms of branchial tubes, characteristic of a fish? For Geoffroy, such observations lent weight to evolutionary ideas. He speculated that higher mammals were pro-grammed to ascend from lower forms of life – their supposed ancestors – to complex forms, during their early development. And to try to prove his theory of evolution, he turned to the fossil record. Cuvier pointed out that if the theory was true, it should be possible to find inter-mediate animal forms in the rocks. If, for example, mammals had evolved from reptiles, where were the intermediate 'reptile–mammal' forms in the fossil record? In effect, Cuvier was challenging him to discover the 'missing links' of evolution.

In 1824, when Cuvier brought out his second edition of *Recherches sur les Ossemens Fossiles*, Geoffroy spotted that he had made an error in his section on ancient crocodiles, and seized his opportunity to score a point over his distinguished rival. From his early travels on the Nile, Geoffroy had become an expert on the anatomy of crocodiles. He quickly recog-nised that Cuvier's ancient 'crocodile' from the Secondary rocks at Caen possessed curious *mammalian* characteristics that Cuvier had overlooked. In particular, the nasal canal in the skull of the 'crocodile' resembled that of mammals. Consequently, he renamed Cuvier's crocodile from Caen, the '*Teleosaurus*', meaning the lizard, or saurian, that announced the approach of mammals in the Tertiary 'from a long distance' (*teleo*) – that is, from the ancient Secondary rocks.

For Geoffroy, such 'transitional' forms were important: the primitive lizards and reptiles had, in some way, given rise to modern creatures. In fact, his detailed observations on the mammalian-like characteristics of the skull of *Teleosaurus* were correct. Cuvier, furious, hurried to reply in 1825 in the *Dictionnaire des Sciences Naturelles*, in which he highlighted the absurdity of a theory in which the Creator made apparently useless creatures merely to fill in the gaps in a Chain of Being. For Cuvier, rather than seeing Nature as one great chain where simple forms give rise to complex forms, believed the animal kingdom should be classified into four distinct groups: the 'vertebrates' such as mammals, fish and reptiles; the 'molluscs' like snails or shellfish; the 'articulates', including insects; and finally, the 'radiates' such as starfish. These four *embranchements*, or branches, were so anatomically distinct that it was inconceivable that one division could have evolved from or developed into another. Of course, Cuvier's ideas were much more acceptable to the theologians; God had surely created the original forms of each group of animals in separate phases of Creation.

In Paris, the debate between these distinguished anatomists did not just echo around the learning centres of the Académie and the Muséum National. Since evolutionary thinking was seized upon by political radicals and linked with reform, it had far greater prominence. As the science historian Adrian Desmond has shown, in the 1820s Republican sympathisers, such as Jean-Baptiste Bory de Saint-Vincent and François Raspail, used evolutionary ideas as a metaphor for social change. After all, if naturalists could show that lower animals through their own efforts could transform themselves into higher animals, could not the down-trodden poor also transform themselves, given a more favourable environment?

Students returning from the Continent to Edinburgh where Richard Owen was studying brought reports of the acrimonious debates going on there. In anatomy classrooms and student societies the radical French ideas were eagerly discussed. At the same time, in Britain the economist the Reverend Thomas Malthus had inspired a radical rebuttal of reform to help the poor. If survival rates improved, he reasoned, the increase in population would outstrip food supply, bringing starvation in its wake. His bleak philosophy was set against a growing chorus in favour of change. Why should not all men gain the vote and empower themselves through education?

As early evolutionary biology became associated with Republican materialism, attacks upon it became increasingly bitter and highly charged. It was condemned by the Church authorities as undermining the moral fabric of society. If Nature was autonomous, where did God fit in? In the words of the *Quarterly Review*, ideas such as these 'break down the best and holiest sanctions of moral obligation and . . . give free reign to the worst passions of the human heart'. If Man could not earn his place in Heaven by his actions on Earth, there was no basis for morality: this would lead to the 'brutalising of man'. Church leaders shuddered at the materialism implied in evolutionary thought. It was unthinkable that life came into being from the 'march of nature' rather than by the Hand of God.

The Reverend William Buckland was one of many leading scientific authorities that did not agree with the evolutionary thinking of Geoffroy. He scoffed at the idea that mammals could have developed from primitive lizards and crocodiles. 'I am not quarrelling or finding fault with a crocodile; a crocodile is a very respectable person in his way,' he teased, 'but I quarrel with finding a man, a *crocodile improved*.' Buckland understood the animal kingdom in terms of the long-held idea of a Chain of Being. For him, the giant reptiles were missing links in God's wondrous Creation. When the Reverend Conybeare outlined details of the ichthyosaur, or 'fish lizard', to his colleagues, he too claimed that it was a link between fishes and reptiles in the great network of being. The *Plesiosaurus*, or 'near to reptile', was a link between the ichthyosaurs and crocodiles. Charles Lyell also wrote to Gideon Mantell when the *Plesiosaurus* was discovered. 'what a leap in the chain we have here, and how many links in the chain will geology have yet to supply'.

Unlike for Lamarck and Geoffroy, for these British experts the rich variety of life shown by each newly discovered 'link' in the chain was not a testimony to evolution. Quite the reverse: God designed each creature for its special place in Nature. In every living thing, a 'guide to God was given'.

In Edinburgh, Richard Owen found all these arguments compelling. At issue was the process of Creation itself: did God make every single species? Was there a series of Creations over time, which would help to make sense of the fact that fossil creatures differed from living animals? Or, as the evolutionists implied, could Nature itself bring new species into being – without God? Owen carefully aligned himself with the mainstream of intellectual thinking in Britain, backed by the Anglican Church. He was drawn to the approach of the celebrated English anatomist John Hunter, whose epitaph celebrated him as 'the gifted interpreter of the Divine Power and Wisdom at work in the Laws of Organic Life'. Owen saw himself as an anatomist revealing God's Wondrous Works, not as an observer of a materialistic process by which Man himself had evolved from lower creatures. In his enthusiasm as a student in Edinburgh he even formed a 'Hunterian Society', where these ideas were discussed.

Within six months, Richard Owen was so highly regarded by his tutor John Barclay that he advised him to develop a career as a surgeon in London rather than complete his university course. At that time, although physicians were required to obtain a degree, those who wished to practise surgery did not have to. It was sufficient for prospective surgeons merely to attend a course of lectures and obtain a certificate from the Worshipful Society of Apothecaries or the Royal College of Surgeons. Barclay recommended that Owen study under John Abernethy, the distinguished President of the Royal College of Surgeons, at St Bartholomew's Hospital.

'I shall never forget the day when I arrived for the first time in London,' Owen wrote of his arrival in April 1825, 'where I literally had not one single friend . . . The sense of desolation which I experienced in walking up Holborn towards St Bartholomew's was something indescribable.' His apprehension increased still further when he caught sight of the great Abernethy engulfed by a group of demanding students and not in the best of moods. Since Owen was very tall it did not take long for Abernethy to become aware of the arrival of a stranger. He turned on him abruptly and demanded, 'And what may *you* want?'

But despite the inauspicious start, Owen thrived in London. His letter of introduction from John Barclay recommended him highly, and he was immediately offered a post as assistant at Abernethy's lectures. To Owen's great advantage, John Abernethy happened to be under fire at the Royal College of Surgeons. The government had paid £15,000 in 1799 for John Hunter's famous anatomical collection. Over ten thousand specimens, which included 'soft animals captured during the circumnavigatory voyage of Captain Cook', had been assigned to the custody of the Royal College. Trustees there were required to prepare them for public exhibition and provide extensive catalogues identifying and explaining the specimens. Twenty-five years on, this work had still not been accomplished. By the mid-1820s, questions were being asked in the press about why it was taking so long before Hunter's great work could be seen. The Royal College was increasingly criticised in medical journals such as the *Lancet* and the *Medical Gazette* as the 'corrupt supporter of a privileged medical elite'. John Abernethy, Sir Astley Cooper and Sir Anthony Carlisle, the senior gentlemen of the College, became a target for the radical press.

Horrified trustees at the Royal College finally learned of the extent of the problem: William Clift, the Curator of the Hunterian Museum, estimated that nine-tenths of Hunter's papers had been stolen by Sir Everard Home, making the task of identifying Hunter's anatomical specimens extremely difficult. It now transpired that Home had destroyed almost all of the original manuscripts. On one occasion, while visiting 'the London Baronet', Clift was mortified to find that Hunter's studies – no less than a national treasure – were being used as toilet paper!

The gentlemen of the College realised they were facing a crisis. They needed someone who could be groomed to be one of them, someone hard-working and dedicated to the work of John Hunter. Above all, they required a talented anatomist who could dissect and identify all the specimens for public exhibition quickly. It was abundantly clear that the ambitious young Richard Owen was their man.

Owen, who received his diploma of membership of the Royal College of Surgeons in 1826, was soon appointed as assistant to Clift at the Hunterian Museum, and gratefully welcomed into the heart of the medical gentry of London. Within two years of arriving in London, he had glided, almost effortlessly, into the privileged inner circle of a wealthy and well connected medical elite.

A letter from his mother at this time shows that she was 'thankful to have a son who has been such a credit to his family'; she points out that 'you are a lucky young man to meet with an appointment of the kind while numbers of the profession hardly know which way to turn'. Catherine Owen, a woman of French descent, took a keen interest in her son's progress, perhaps all the more now that her only other son, James, had died. She warned Richard to take great care during dissections and 'to make a point of washing your face, neck, arms and hands &c every night before you enter bed'. Shrewdly she told him, 'one thing indeed that can never be too strongly recommended to young men aspiring to rise in their profession . . . is to become pupils of some person already eminent and in high repute; by such a course they obtain two great objects – a well grounded professional knowledge and the opportunity of becoming known to all the friends and connections of their instructor'.

The diligent young Owen did not restrict himself to mere acquaintance with the friends and colleagues of William Clift at the Royal College. It wasn't long before another opportunity arose. Clift's only daughter, Caroline, had been hanging a pair of bell-pulls in her mother's room that she had made for a present. As she gathered her skirts to come down the step-ladder she fell awkwardly, and collapsed with minor injuries. Faced with this emergency, with the Clifts' only daughter apparently insensible, all notions of social etiquette were thrown to one side. The nearest doctor to hand was hastily summoned. Caroline was laid on a chaise longue and given smelling-salts, and as she opened her eyes she found herself face to face with Richard Owen, studiously attending to her injuries. Perhaps disarmed by the intimacy of the situation or the unexpected charm of the man, Caroline Clift found it difficult not to notice her father's young apprentice, and soon Richard Owen was beginning to show more than polite attention to his master's daughter.

Caroline Clift's mother had very high aspirations for her daughter. A man of considerable income or property was the very least that she hoped to achieve. A junior at the museum with no accumulated wealth was definitely not her ideal son-in-law. Under normal circumstances, Richard Owen would not have moved in the same circles as Caroline Clift.

But despite the fact that he was considered an inappropriate suitor, Owen continued to court her. Caroline began to write about him in her diary: 'R.O. gave me a carved tortoiseshell comb', and 'R.O. gave me a volume of Cowper's poems.' Presumably at her insistence, Owen was now invited to the house on social occasions to attend musical evenings and dinner parties. To the dismay of his future mother-in-law, Richard and Caroline were engaged within three months. By Christmas 1827, at the age of twenty-three, Richard Owen had positioned himself socially and professionally at the very centre of the scientific stage in London.

In the museum at the Royal College he applied himself with tremendous energy to restoring John Hunter's collection. Surrounded by large bottles of preservative alcohol and literally thousands of anatomical specimens in various stages of decay, he set to work on the monumental task of preparing Hunter's works for exhibition. He rapidly won over his future father-in-law, who was greatly relieved to have such an able and enthusiastic assistant. William Clift described him as 'sober and sedate – very far beyond any young man I ever knew'.

Impressing the amiable Clift was not Richard Owen's sole object. He soon learned that Baron Georges Cuvier intended to visit the Royal College museum to update his research on fossil fishes. And it became apparent that no one in the museum could speak French, except Clift's studious assistant. Owen may not have applied himself in school, but, thanks to his French mother, he was in possession of that extra skill that could serve as his passport to the higher echelons of Parisian intellectual society. He was the obvious person to serve as host to the distinguished Baron.

They were a strange sight: the earnest young Owen, gangly and tall, escorting the sixty-one-year-old Baron, now hugely fat and unkindly nicknamed 'the Mammoth'. They walked around the museum at a slow, regal pace in keeping with both the Baron's dignity and his physical limitations. Still encumbered with an eager impatience and annoyance at those who did not understand him, for the Baron to find an admiring Englishman who not only understood anatomy but also could talk as fluently as a Frenchman was refreshing indeed. As for Richard Owen, to be seen conversing on intimate terms with the great Cuvier, explaining the exhibits in another language that made all others redundant, was a victory to be savoured. To his great delight, Cuvier invited him to visit the Muséum National the following year.

On his arrival in Paris the next summer, as he was alighting from his carriage, Owen was told of illuminations and fireworks that were about to take place. Jostled by the crowds and narrowly avoiding the carts, he hurried to the Place Louis XV, and 'on a sudden turn, found myself in the most beautiful place in the world, amongst the noble walks, statues, fountains, glassy pools . . . of the Jardin des Tuileries . . . A gun fired and a rocket shot up from the front of the Palace, which was answered by an immense flight of rockets, red, green and blue balls from the Pont de la Révolution . . . then came a shower of lights that made it brilliant daylight.' As he watched the spectacular display he could not fail to be thrilled. It was an auspicious start to his visit.

The next day, he was ushered into Baron Georges Cuvier's inner sanctum at the Muséum National d'Histoire Naturelle. Owen described the work in progress on Hunter's collection and deferentially passed on William Clift's respects. Cuvier personally showed him around the museum and urged him to visit whenever he pleased. During his several months in Paris, Owen became a welcome visitor at the Baron's famous Saturday evening soirées. He attended Cuvier's lectures and even worked in the dissecting rooms and public galleries of the museum alongside the great man's assistants.

The extensive collection of fossils in Paris made a lasting impression on Owen. He became aware that behind Cuvier's great success was this extraordinary museum, the best of its kind in the world: a great cathedral filled with the trophies of the natural world, whose airy corridors and high-vaulted ceilings were crammed with the most remarkable creatures known to science. There was nothing comparable to this in England — yet.

His mother followed developments in Owen's career with delight: 'Your being noticed by Cuvier was fortunate indeed,' wrote Catherine Owen, 'and your having access to his museum would be an advantage in your profession on many accounts, and I trust you will reap the benefit of it ultimately.' Already, Owen had positioned himself at the forefront of anatomical research and taken lessons from the great Baron himself. The northern boy who had been forecast to 'come to a bad end' was doing rather well, and this was only the beginning.

7

Satan's Creatures

The Fiend
O'er bog or steep, through strait, rough, dense, or rare,
With head, hands, wings, or feet pursues his way,
And swims or sinks, or wades, or creeps, or flies.

Milton, *Paradise Lost*, Book II

While Richard Owen was immersed in preparing Hunter's catalogues, familiarising himself with the anatomy of living animals, geologists continued to expose ever more tantalising clues to the ancient creatures. It had long been known that among the more curious remains embedded on the shore at Lyme Regis were numerous dark-grey stones, like oblong pebbles or kidney potatoes and marked with distinctive whorls. Initially they were called 'Bezoar stones' following William Buckland's observation, to his somewhat bemused colleagues at the Geological Society, that they resembled 'the concretions in the gall-bladder of the Bezoar goat, once so celebrated in medicine'. The experts were quite baffled as to how these extraordinary artefacts fitted into the ancient world.

At first, Buckland thought that the curious ripples on the surface of the bezoars had been created by the action of waves on soft clays. Gideon Mantell had found similar fossils with marked whorls in the sandstone of the Tilgate Forest. He discussed them with Buckland, and they realised that the bezoars were more frequently found in strata that were rich in animal fossils. Buckland was already familiar with the petrified faeces of hyenas from his study of the cave in Kirkdale, and it occurred to him that the excrement of ancient sea lizards embedded in the older Secondary rock at Lyme could also have been fossilised. This opinion was further substantiated by Mary Anning, who noticed the bezoars were common *inside* the ichthyosaurs, usually in the abdomen near the rib-cage or pelvis. It became increasingly evident that these were, indeed, fossil faeces, the spiral markings formed by their passage through the intestines.

The stones, which soon became known as 'coprolites', revealed the partially digested remains of food that had been devoured by the monsters of the ancient world. 'Dispersed irregularly and abundantly throughout these petrified faeces are the scales, and occasionally the teeth and bones of fishes,' observed the Reverend Buckland, 'that seem to have passed undigested through the bodies of the Saurians.' He sent some coprolites for chemical analysis to Dr William Wollaston, a friend at the Geological Society. Wollaston soon reported that the specimens were rich in phosphate of lime, suggesting that a large proportion of the Saurian diet had been the bones of other creatures.

Within geological circles in Georgian England the new science of coprology was somewhat hampered, since it was deemed too indelicate to discuss. 'Though such matters may be interesting,' Wollaston warned Buckland, 'it may be as well for you and me not to have the reputation of too frequently and too minutely examining faecal products.' Buckland, however, jumped right into the study of 'coprology' with his usual abundant enthusiasm. 'When we see the body of an Ichthyosaurus still containing the food it had eaten just before its death,' he said, 'ten thousand, or more than ten thousand times ten thousand years ago, all these vast intervals seem annihilated, come together, disappear, and we are almost brought into as immediate contact with events of immeasurably distant periods as with the affairs of yesterday.' He spent weeks with Mary Anning studying the many bezoar specimens found at Lyme. Although most coprolites were four to six inches long, 'some are much larger, and bear a due proportion to the gigantic calibre of the largest Ichthyosauri,' observed the professor. 'Some are flat and amorphous, as if the substance had been voided in a semi-fluid state, others are flattened by pressure of the shale.'

William Conybeare informed him that similar bezoars could be found near Bristol, and soon Buckland was travelling the length and breadth of the country trying to find fossil faeces in their different formations. At Oxford, his latest academic pursuit couldn't fail to attract attention, and it even inspired some of Buckland's colleagues, including one John Shute Duncan, to poetry – of a sort:

> Approach, approach ingenuous youth
> And learn this fundamental truth
> The noble science of geology
> Is firmly bottomed on Coprology

Buckland's research revealed a disturbing image of the ancient oceans, with sea lizards locked in an unending and vicious carnivorous combat. Some coprolites contained the undigested vertebrae and teeth of small

'Satan's creatures': the ichthyosaur fighting the plesiosaur
from Figuier's *The World before the Deluge*, 1866.

ichthyosaurs. 'These monsters of the deep may have devoured the smaller and weaker of their own species,' explained Buckland. Miss Anning, he remarked, had found two coprolites together, close to the skeleton of an *Ichthyosaurus*, 'as if they had been voided by it in the struggles of death'.

For Buckland, the coprolites were proof of 'records of warfare, waged by successive generations of inhabitants of our planet on one another . . . The general law of Nature which bids all to eat and be eaten in turn is shown to have been co-extensive with animal existence on our globe.' This former creation in which such horrific carnage was part of daily life was inexplicable. The sheer ungodliness of these cannibalistic sea creatures, with their smiling jaws and monster appetites, beggared belief. As the geological enthusiast and amateur collector Thomas Hawkins pointed out, these were more like Satan's creatures: 'armed with all the virility of Evil, instant sprung, Satan seems to have been thrice seized, generating Horrors and realising a teeming Spawn fitted for the lowest abysm of Chaos!' Why would God create such hideous monsters?

If the idea of an ancient ocean filled with Satan's creatures was perplexing enough, attempts to reconstruct life on land proved even more unsuccessful. In Lyme Regis there were entire skeletons that showed what the sea lizards were like and now even fossil faeces that graphically revealed their colourful eating habits. But in the Tilgate Forest in Sussex there were no complete skeletons; every fragment of fossil evidence was scattered and buried within layer upon layer of rock.

After his success at the Royal Society, Mantell embarked on a second book in which he was determined to provide a complete survey of the Tilgate Forest of the Weald. To help him interpret the fossil evidence, when he found time from his practice he extended his network of correspondents. Apart from his familiar acquaintances such as the Wiltshire collector Etheldred Benett, he was writing to members of the Geological Society such as Charles Lyell and William Fitton, and to Davies Gilbert at the Royal Society. He also established foreign correspondents 'of the first order', including Georges Cuvier and his colleague at the national museum in Paris, Adolphe Brongniart, pioneer in fossil botany.

By the mid 1820s, Adolphe Brongniart's research on fossil plants was revealing tantalising glimpses of a history of plant life on the land. The oldest rocks of all, the Primary rocks, were totally devoid of any signs of life. Above these in the uppermost layers of the ancient Transition rock, Brongniart identified simple plants showing little variety, such as mosses. The first conifers made an appearance towards the end of the Transition series. Above this, in the highly stratified Secondary rock formations where reptiles were being uncovered, tree-ferns and conifers dominated the vegetation and vast forests flourished. Above this level in the geological sequence, in the Tertiary rocks, he found flowering plants.

The history of plant life, said Brongniart, was one of increasing complexity of forms. Some of the oldest and most primitive forms, including ferns, horsetails and mosses, reproduced simply by means of spores. After this, the gymnosperms such as conifers appeared; these had big naked seeds in their cones. Last of all to make their appearance in the fossil record were the flowering plants known as angiosperms. These are more complex in that they have male and female seeds and can repro-duce sexually, with wind or insects disseminating their seeds and fruits. Brongniart's research showed that plant life had changed over time; it was, in some way, progressive.

He linked these changes in plant life to different conditions on the surface of the earth. His research into fossil plants buried in the coal measures, low down in the Secondary series of rocks, confirmed evi-dence produced by Mantell, Buckland and many others that the ancient globe was much hotter than the present. The giant tree-ferns and club-mosses that Brongniart found in the coal measures bore most resemblance to plants from lush, tropical rain-forests. He concluded that early in the history of the globe there had been hotter conditions and a smaller variety of plant life. Much later, the earth cooled and plant life diversified.

Gideon Mantell sent fossils to Brongniart in Paris, who helped him identify more ancient plants from the Secondary strata of the Weald. Mantell had found intriguing oblong fossils like fruits, which he thought might be palm kernels. Embedded with them, criss-crossing the stone in a network of lines, were the impressions of segmented stems. With Brongniart's help, the 'fruits' were identified as the tubers of horsetails, or *Equisetum*. One species that Mantell found in great abundance he named after his friend: *Equisetum lyellii*.

Apart from giant horsetails with their tall, spiky stems and tubers, there were conifers, cycads and above all, ferns in such profusion that, Mantell observed, 'no considerable block of stone is without traces of them'. Some of the ferns, he thought, belonged to varieties that grew to only three or four feet, while the tree-ferns could reach forty feet. The fronds of the ferns, he observed, 'appear to be very distinct from those of recent ferns. This is yet another instance of the difference existing between the vegetation of the ancient and modern condition of the planet.' The earth of antiquity, with its high temperatures and strange vegetation, was, he thought, 'wholly different from its present state' and was 'probably unfit for the habitation of creatures of a more perfect organisation'.

By 1826, from dental evidence alone, Mantell suspected that at least *four* different species of reptiles had once thrived in the tropical forest that had covered the ancient Weald: crocodiles, plesiosaurs, *Megalosaurus* and *Iguanodon*. The teeth, he said, in both 'form and structure present such striking differences as to be readily distinguished from each other and from those of existing species'. But there the trail to reconstruct the ancient skeletons always stopped. It was very hard with any scientific honesty to classify the bones that he had found. 'The fossil bones possess so many characters in common, that when we considered the broken and detached state in which they occur and their intermixture with the debris of turtles, vegetables fishes and shells,' Mantell wrote, 'the difficulty of the attempt to identify the bones of the respective animals seems almost insurmountable.'

Some of the bones that Buckland appropriated to *Megalosaurus*, Mantell thought possibly belonged to one of the other species. Once,

The popular impression of *Megalosaurus*, featured in
the *Penny Magazine*, 1833.

when the Reverend Conybeare visited Lewes, they went through the
entire collection of vertebrae, separating out those that might have
belonged to the crocodile, *Plesiosaurus* and *Megalosaurus*. Having done so,
several enormous vertebrae remained, some ten inches wide. 'From
their enormous magnitude,' Mantell wrote, 'we are induced to refer
them to the Iguanodon, the only other gigantic lizard of the Tilgate
strata.' But without a complete skeleton, or even a few connected
bones, there was no proof that his inferences were correct. With these
scraps of evidence, he was obliged to classify the bones under vague
headings such as 'Bones whose characters are not determined' or 'Bones,
supposed to be referable to the Iguanodon'.

As for discerning the heads of the giant beasts, without the bones of
the skull it was all sheer conjecture. Since the *Iguanodon* teeth were
herbivorous, the only inference he could make with certainty was that
the reptile's head was different from anything known. 'It follows
from the peculiar structure of the fossil teeth alone,' he reasoned, 'that
the muscles which moved the jaws and the bones to which they were

attached, were widely different from those of any of the living lizards and consequently the form of the head of the Iguanodon must have . . . differed from those of existing reptiles.' But how to picture the muscles of the face, the soft tissue of the cheeks, the tongue and lips? It was utterly baffling.

The only other clue to the shape of *Iguanodon* that Mantell possessed was the horn – that curious appendage originally cited as evidence that the unknown herbivore was a rhinoceros. When he returned to the Hunterian Museum in 1825, he discovered that there was a species of iguana that had a horn on the forehead, rather like a modest unicorn. 'Between the eyes and nostrils, are seated four rather large scaly, tubercles; behind which rises an osseous conical horn,' he observed. 'That our fossil was such an appendage there can be no doubt.'

Even though the evidence on their anatomy was so sparse, the size and majesty of what had once been living beasts always held him in thrall. The thigh bone of *Iguanodon*, he wrote, was so 'monstrous . . . [that] were it clothed with muscles and integuments of suitable proportions, where is the living animal, with a thigh that could rival this extremity of a lizard of the primitive ages of the world?' Even the metatarsals, the bones in the sole of the foot between the ankle and the digits, were 'so large that they appear more like the bones of mammoths or elephants than of reptiles'. One of these metatarsal fragments was more than four inches long and thirteen inches in circumference. 'Were we to calculate the probable magnitude of the original animal from the data which this metatarsal bone affords,' he remarked, 'our readers might well exclaim that the realities of Geology far exceed the fictions of romance!'

As if to add to the far-fetched image of this ancient world, he was struck by another bizarre observation. 'The organic remains of the Tilgate Forest are exceedingly numerous,' he wrote in 1826. In every pit he investigated at Tilgate the lower layers of bluish-grey sandstone had compacted into conglomerate. This 'conglomerate' was so rich in animal fossils that the larger pebbles seemed to be composed principally of fragments of bones and sandstone – even the finer parts were formed of ground-up bones, teeth and sand. The lonely stone quarries which now echoed to the sound of his chisel told of a time when ancient life was *prolific*.

Although this ever-present ancient world claimed his attention, dominated his mental and imaginative life, the endless commitments of Mantell's practice had to come first. He kept up to date with the latest medical developments, and was one of the first to use the new drug 'ergot of rye' in cases of prolonged labour, publishing his successful results in the *London Medical Gazette*. At this point he had attended almost two and a half thousand deliveries. And for the poor he was still the first port of call for countless gruesome medical emergencies in the district: 'Performed an operation on Funnell's boy; the removal of a very large "fungus haematodes" from the back of the head; nearly two pounds in weight,' he recorded. On another occasion, 'with Hey's saw, removed several portions of skull that had been forced into the brain. A boy, 16 years old crushed by a horse, died next morning.' Rather more successfully, he dealt with a 'Mr Weller of Southover who met with a frightful accident; from the discharge of a gun loaded with shot; he shattered his left arm, wounded his thigh most severely and injured his face and eyes. I amputated the forearm in the presence of Mr Hodson who was pleased with my dexterity; at present the case is going on well.'

And he still made time to intervene in cases of social justice. In 1826 a poor woman, Hannah Russell, and her lodger were sentenced to death by hanging, accused of murdering her husband. He had collapsed suddenly and died while stealing a sackful of corn from a neighbouring farm in the middle of the night. It was thought that Hannah and her lodger were having an affair and had conspired together to poison him with arsenic. Witnesses argued that Hannah Russell had bought 'poison' at the village shop; another witness insisted that she had been seen spreading a white powder on a slice of bread and butter, which Hannah had claimed was for the mice. Mantell, who had entered the courtroom purely by chance, was concerned at the scanty medical evidence on which the death sentences were imposed. He carried out more investigations and proved that her husband had in fact died of heart disease, having suffered from angina for a few years. His evidence was compelling, and as a result Hannah Russell was given a free pardon. Unfortunately, the pardon was too late for the young lodger; within a week of the first sentence, he had already been executed.

Against this burden of responsibilities, the research for his second book on the Tilgate Forest was continually delayed and he had little time to write to potential subscribers. It is likely, too, that from the late 1820s his wife Mary was putting him under considerable pressure to give up geology and concentrate on medicine. 'This is the last volume, in all probability that I shall ever publish on geology,' Mantell wrote as he prepared the introduction to his book. 'Interesting and numerous as are the relics of a former world which my humble labours have brought to light they are, in truth, but mere indications of the vast and important geological treasures which remain to reward more active, judicious and extended researches than mine.'

Since the arrival of his third child, Hannah Matilda, in November 1822, Mantell's family responsibilities were only too clear. He was very fond of his three children. Sometimes they would accompany him, exploring local sites and quarries, rambling on the beach in search of fossils, and occasionally they joined their father on trips to London. Hannah was renowned for her charm and her gentle nature, and it was quite clear that she was his favourite: he always referred to her in his diary as 'my darling child Hannah' or 'my sweet Hannah'.

Mantell's *Illustrations of the Geology of Sussex* was finally published in January 1827. In it he outlined the years of painstaking research spent classifying the Sussex rocks. Almost half the book was devoted to summarising the details of the saurian animals he had found, which could now be correctly named and identified. He confirmed that at least four types of giant saurian had once thrived in the Weald: ancient species of crocodiles, *Megalosaurus*, *Iguanodon* and *Plesiosaurus*. He described how these creatures were found, attempted to assign bones to each, and discussed the difficulties involved in such classification.

As he strove to convey his excitement, he built up a vivid picture of the ancient world.

Consider an estuary, formed by a mighty river flowing, in a tropical climate, over sandstone rocks . . . through a country clothed with palms, arborescent ferns . . . and inhabited by turtles, crocodiles and other amphibious reptiles . . . The gigantic Megalosaurus and the yet more gigantic Iguanodon, to whom the groves of palms and arborescent ferns would be mere beds of reeds, must have been of such prodigious magnitude that the existing animal creation present us with no fit objects of comparison. Imagine an animal of the lizard tribe, three or four times as large as the largest crocodile, having jaws, with teeth equal in size to the incisors of the rhinoceros and crested with horns; such a creature must have been the Iguanodon! Nor were the inhabitants of the waters much less wonderful; witness the Plesiosaurus which only required wings to be a flying dragon.

This, the first publication dealing principally with giant fossil reptiles, has recently been described by Mantell's biographer Dennis Dean as 'the rarest and most historic dinosaur book in English'. But in 1827, as the months wore on, hardly any interest could be discerned. This may have been partly due to the fact that Mantell's discoveries were still presented in the context of the local geology of Sussex, as shown by the book's lengthy sub-title, *A general view of the geological relations of the South-Eastern part of England with figures and descriptions of the Fossils of the Tilgate Forest*. The significance and universality of the beasts he was describing were not yet recognised.

In addition, the book was costly for the publishers, Lupton Relfe, to produce. Only a hundred and fifty copies were ever printed, and sales were slow. For Mantell, this failure after all his Herculean efforts, combined with his wife's continued protests over his commitment to geology, was beginning to put a strain on his marriage.

Shortly after the birth of their fourth child, Reginald Neville, in August 1827, Mary Mantell began to spend longer periods away from her husband. 'So many vexations have beset me,' Mantell confided in his diary in the September, 'that I have not had the inclination to record days of wretchedness, the disappointment of every long cherished hope.' By the autumn it was becoming crystal-clear that this book was even more of a financial failure than his first. Only fifty were ever sold. Just before Christmas Mary went to stay with her own family and remained away for weeks. 'Gracious heaven, Oh relieve me from the pangs I now suffer,' wrote Mantell.

A year later, a startling new discovery on the shore at Lyme drew attention once more to the strange lost world of reptiles. In December 1828, Mary Anning uncovered the fragile remains of an unearthly creature from a distant time that seemed part vampire-bat, part reptile, bearing no resemblance to anything she had seen before. Unlike the giant sea lizards, the ghostly relic that emerged from its limestone grave on the

Reptiles Restored by George Scharf, showing, on the left, Mantell's conception of the *Iguanodon* as a giant lizard.

shore was slight and misshapen, the fine elongated bones dislocated, compressed over the millennia at awkward angles to each other. The head was missing entirely.

Like some bird of prey, not much larger than a raven, this flimsy assortment of fragments of bone, hints of claw and wing, was an unexpected new revelation. Fragments of hollow, light, bird-like bones had been reported before at Lyme, but their identity had not been confirmed. William Buckland hurried to see the evidence: 'Miss Mary Anning . . . has recently found the skeleton of an unknown species of that most rare and curious of all reptiles,' he declared, 'which has yet only been recognised in the limestone beds of Aichstedt and Solnhofen: the Pterodactylus.'

Fifty years earlier, an Italian naturalist, Cosmo Alessandro Collini, had uncovered the fragmentary remains of a bird-like creature in a quarry near Solnhofen in Bavaria. Collini had placed the strange fossil in the Grand Ducal Museum at Mannheim, where it was identified as an ancient aquatic animal. Others maintained it could be a bird, a bat, perhaps even a new kind of vampire-like beast. In 1809, Georges Cuvier had also studied the Bavarian fossil and concluded from the structure of the jaw and skull that this was a reptile. He had been intrigued by the fourth finger of the animal, which was greatly elongated. He had speculated that this finger had once supported a wing. Indeed, the lightness of the bone and its configuration suggested this odd creature could fly. Of all the beings in the ancient world, he had declared, this was 'incontestably the most extraordinary': it was a *flying reptile*. He had named it the '*Pterodactylus*', or 'wing finger'.

Cuvier's highly original observations were disputed. Some German naturalists, such as Johannes Wagler, still believed that the *Pterodactylus* was aquatic. Wagler thought it had similarities to the skeleton of the ichthyosaur, or sea lizard, and tried to reconstruct the *Pterodactylus* fossils as though the creature moved like a swan in the water, its long arms serving as flippers, like a penguin's. Although Mary Anning's new creature, named *Pterodactylus macronyx* by Buckland, was more perfectly preserved than the earlier fossils, the uncertainty over the nature of the beast persisted. The matter was only resolved ten years later, when it was observed that *Pterodactylus* bones had air ducts to lighten the skeleton. This pointed to Cuvier's conclusion: the reptile was created for flight.

The flying reptile, the first to be identified in England, helped to revive the Annings' flagging fortunes. They had again hit on hard times, since demand for the giant sea lizards was slacking and many of their patrons had suffered their own financial crises. Richard Grenville, 1st Duke of Buckingham, who had previously bought several of the Annings' fossils, found himself in such straitened circumstances by the mid-1820s that he could no longer bid for them. It transpired that the Duke had squandered much of his wealth on luxurious entertainment and purchases of 'fine works'. In order to escape the not inconsiderable embarrassment for an aristocrat of being pursued by creditors, he often found it convenient to spend long periods away on his yacht. The Annings had also been patronised by the Bristol Institution, but this, too, was facing a financial crisis since several local banks had collapsed after damaging speculation in the London markets.

Since even the British Museum was unable to find the funds to buy the *Pterodactylus*, William Buckland bought the new creature himself. At the Geological Society, he described it in vivid terms to his colleagues:

> The Pterodactylus somewhat resembled our modern bats and vampires, but had its beak elongated like the bill of the wood-cock and was armed with teeth like the snout of a crocodile; its vertebrae, ribs, pelvis, legs and feet resembled those of a lizard; its three anterior fingers terminated in long hooked claws like that on the fore-finger of a bat; and over its body was a covering . . . of scaly armour like that of an Iguana; in short it was a monster resembling nothing that has ever been seen or heard-of upon earth excepting the dragons of romance and heraldry.

He explained that the wings, when unfolded, must have reached a span of four feet. The other three fingers terminated in very long paws with

Duria Antiquior or *Ancient Dorsetshire* by Henry de la Beche,
1830, portraying the carnivorous warfare of the ancient ocean.

which the creature could creep or climb or suspend itself from trees.
'Thus like Milton's Fiend,' declared the Reverend Buckland, 'this
strange tenant of the infant world "With head, hands, wings, or feet
pursues his way,/And swims or sinks, or wades, or creeps, or flies."'

These vampire-like flying-reptiles with their huge eyes and sharp
teeth and claws, captured the public's imagination. 'They were nature's
first attempts at anything in the bird line,' Charles Dickens wrote later.
'Even if pre-Adamite man is ever proved to have been existing at that
time, we cannot imagine his wife making pets of them, or his children
liking to have them hung about the house in cages; they have such a
family likeness to the Evil Spirits who beset Aeneas or Satan in an old
illustrated Virgil or *Paradise Lost*.' Inspired by Mary Anning's finds, the
geologist Henry de la Beche produced an illustration of ancient Dorset,
Duria Antiquior, depicting an ancient ocean filled with predatory

plesiosaurs and ichthyosaurs, locked in combat, while pterodactyls circled above.

The new discovery brought sharply into focus the fantastic variety of reptilian life found in Secondary formations throughout Europe. By 1828, pterodactyls had been uncovered in the Bavarian limestones on the Continent and in the beds at Lyme and, Buckland speculated, the bones of birds found in Stonesfield also might belong to pterodactyls. Plesiosaurs, mosasaurs, ichthyosaurs and crocodiles had come to light in France and southern England. The reptilian fossils Mantell had found in the Weald corresponded to fossils at other sites of different ages in the Secondary series, such as the Stonesfield slate and the blue lias of Lyme.

Gideon Mantell followed each development with great interest. His wife had returned to him and they had renewed their efforts to repair any misunderstandings; she had even come with him to the quarry, an outing that Mantell described as 'glorious'. Unlike some of the gentlemen scholars in London, Mantell was acutely aware that the astonishing beasts that were reported were not isolated examples. From his almost daily forays into the quarries of Sussex which he would not give up – even for his wife – he knew that reptilian remains were abundant. 'Some of the reptiles, from their organization, have been fitted to live in the sea only,' he observed, 'while others were terrestrial, and many were inhabitants of rivers and lakes.' Now there was evidence of reptiles in the air.

How did this fit with Cuvier's 'Age of Mammals' in the more recent Tertiary strata, which lay above the Secondary rocks in which the reptiles were buried? A jaw from a mammal, an opossum-like creature, had been found in the ancient rock at Stonesfield, but apart from this, there were no mammals in the Secondary rocks. A distinct order was beginning to emerge in the fossil record of animal life on the planet, much as Adolphe Brongniart had revealed in the case of plant life. Mantell wrote:

The prodigious quantity of the remains of these reptiles which has, within a comparatively short period been found in England alone, is truly astonishing. If to these we add the immense numbers that have been discovered in France, Germany &c and reflect that for one individual found in a fossil state, thousands more must have been devoured or decomposed; and that even of those that are fossilised, the number that comes under the notice of the naturalist must be trifling compared with the quantities unobserved or destroyed by labourers, we shall have a faintest idea of the myriads of 'creeping things' which inhabited the ancient world.

The 'carnivorous carnage' revealed by the coprolites, the sheer number of animal bones unearthed in the Secondary rock of the Weald, and the variety of reptilian life – all this, for Mantell, pointed to a bizarre conclusion. There was a period in the earth's history when 'creeping things' dominated the landscape. But if so, what was that landscape like and what could have happened to those bizarre and prolific forms of life? How had they been so totally wiped from the face of the earth, only teeth and bones remained?

8

The Geological Age of Reptiles

Art, Empire, Earth itself, to change are doomed;
Earthquakes have raised to heaven the humble vale,
And gulfs the mountain's mighty mass entombed,
And where the Atlantic rolls wide continents have bloomed.
Cited in *Thoughts on a Pebble* by Gideon Mantell, 1849

While Gideon Mantell was puzzling over the evidence from the Weald, Charles Lyell was working on radical ideas that would help to lay the foundations of geology and shed new light on how the animals of antiquity became extinct. Abandoning his career at the bar, in the spring of 1828 he set off for the Continent with his friend Roderick Impey Murchison. Murchison had originally embarked on a career in the military but, like Lyell, had sufficient wealth to pursue his fascination for geology. Accompanied by Mrs Murchison, they toured central France studying the landscape.

As their open carriage wound its slow and measured way through the French countryside, revealing its contours in intimate detail, it became increasingly obvious to Lyell that Buckland's theories of a biblical Deluge did not fit, and especially after their detailed study of rivers in the Auvergne. For Buckland, valleys were excavated by a single dramatic event, a great cataclysm that flooded the world. At first sight, some of the spectacular valleys in the Auvergne, with their modest streams winding their way through, appeared to support Buckland's view. The sizes of the streams and of the valleys were unrelated. But Charles Lyell, building on earlier observations by an amateur geologist called George Scrope, believed these formations had nothing to do with a sudden Deluge.

Lyell wondered whether the valleys were formed gradually, as rivers in this volcanic district carved their paths through successive layers of lava. He was struck by the excavating power of streams that had been blocked up by solid currents of lava. In these cases, he could find no evidence 'that the sea, nor any denuding wave, or extraordinary body of water, have passed over the spot since the lava consolidated'. The streams themselves had repeatedly re-excavated a passage through a series of lava flows, sometimes to a great depth.

As he extended his observations, Lyell found a site where gravels – which Buckland would have maintained were shifted by the waters of the Flood – lay beneath a volcanic deposit in which a valley had formed. It was hard to explain how a single event, like the biblical Flood, could have deposited the gravels and simultaneously formed a valley in strata of different ages lying above them. This could only fit with Buckland's theories if there had been more than one Deluge. Lyell therefore concluded that the valleys in the Auvergne were not formed by the biblical Flood, but by erosion from local rivers.

William Buckland was riding a wave of popular appeal in England and was busy with his second volume of '*Reliquiae Diluvianae*, or Relics of the Deluge'. When he heard the evidence that Lyell and Murchison had uncovered in France, he quietly abandoned his plans to publish. Among members of the Geological Society, belief in Noah's Flood was increasingly becoming an article of faith, as immortalised in a rhyme by Shuttleworth: 'Some doubts were once expressed about the Flood,/ Buckland arose, and all was clear as – mud!' William Conybeare, the parson-geologist, rallied to his friend's defence. George Scrope, the naturalist who had influenced Lyell, was a mere 'goose', he reassured Buckland. Conybeare drew attention to English valleys that had no streams winding through them, and huge boulders dumped on the valley floor. These valleys did not appear to be formed by rivers or by any other natural cause, but could indeed have been created during a great Flood. He now embarked on a detailed memoir examining valley formation.

In retaliation, Charles Lyell was determined to 'free the science from Moses'. England, he declared, was 'more parson-ridden than any country in Europe except Spain', and this was detrimental to scientific thinking. When Roderick Murchison returned to London to recover from a fever, Lyell continued his tour alone into Italy and Sicily, relentlessly seeking out any evidence that would remove the hand of God from the features of the landscape. He became immersed in the study of every natural process that could conceivably be responsible for altering the earth's surface: the erosion caused by streams, rivers and waves; the gradual accumulation of sediments in deltas and on the ocean floor; the uplift of strata due to earthquakes and volcanoes. These, Lyell believed, were the 'alphabet and grammar of geology', which could create the 'humble vale' and raise the 'mighty mountain'. Imperceptibly, over millions of years, these processes could explain great changes in the earth. There was no need to invoke a Flood or a series of Cuvierian 'catastrophes'.

Lyell argued that the present was the key to the past. All the geological processes that could be studied today had occurred throughout history and at the same rate. Processes such as erosion or the accumulation of sediment are so slow that they are barely noticeable, and have always been so. It was only because of gaps in the fossil record that changes in the past appeared to occur with dramatic speed. Events separated by a great distance of time seemed to follow each other abruptly, lending weight to Cuvier's catastrophes or violent revolutions. In fact, the surface of the earth had always been subject to slow and steady change.

Georges Cuvier had used his theory of catastrophes to explain extinction: 'numberless living beings', he claimed, had been destroyed by 'terrible events'. Lyell provided a different explanation of extinction. Species could only survive if conditions in their environment favourable to their existence remained the same. Since Lyell had shown that the living world was constantly changing, with geological processes imperceptibly altering the landscape all the time, habitats were sometimes wiped out and species lost. Each alteration in a habitat would affect many species, favouring some and destroying others, depending on the complex relationships each animal species had with others. In Lyell's view, ordinary geological processes could explain great changes in the history of life. For the first time, Divine intervention was convincingly removed from the earth's history.

When Lyell returned from his tour in January 1829, he began to prepare his ideas for a book that he was planning, *The Principles of Geology*. Although immersed in his work, in March of that year he took the trouble to write to his friend Gideon Mantell. Aware of the difficulties Mantell faced in trying to establish a geological career while coping with a busy medical practice, in his letter he insisted that Mantell should make his name shine in the scientific world by seizing the opportunity to take the lead in the field of fossil reptiles.

> Now I have sworn with myself that you shall show them ere many years who and what you are and put to blush the jealous unwillingness which most metropolitan monopolists in Science, both in France and here, exhibit towards all such as happen not to breathe their own exclusive atmosphere. But you must concentrate . . . Cut out botany . . . give up all ideas of a popular book on geology . . . But from this moment resolve to bring out a general work on the subject of 'British Fossil Reptiles and Fish' . . . Clift has no time; Buckland is divided amongst a hundred things and is no anatomist . . . You must say nothing for a while. By this you may render yourself truly *great* . . . the field is yours but might not remain open many years. It is worthy of your ambition and the only one which in an equally short time you could make your own in England and for ever.

Gideon Mantell listened to his friend's advice. Later that month he took steps to reduce his medical workload, signing an agreement with a young doctor, George Rickward, who wished to buy a share in his Lewes practice. With more free time, he immediately set to work enlarging his collection of saurians. Rather than displaying fossils just from Sussex, he wanted key specimens from other important sites. Buckland, Murchison and other friends sent donations; each day brought new messages from the Lewes waggon office heralding the arrival of another hamper.

In April, a large box of seventy casts arrived from Georges Cuvier. With great disruption to the household, a large new room was built on, to house the extended collection. In June, carpenters took instructions on the design of many new cabinets. As the work progressed, Mantell was so preoccupied in Lewes that he was unable to attend Geological Society meetings in London. Lyell's letters were a welcome diversion that kept him informed of heated exchanges, as Conybeare and Buckland were forced to retreat from the idea of a Deluge. Lyell wrote that same month:

> My dear Mantell – the last discharge of Conybeare's artillery drew upon them on Friday a sharp volley of musketry on all sides as was enough to sink Buckland's *Reliquiae Diluvianae* for ever, and make the second volume shy of venturing out to sea . . . Murchison and I fought stoutly. Buckland was very piano. Conybeare now admits three deluges before the Noachian! And Buckland adds God knows how many catastrophes besides – so we have driven them out of the Mosaic record fairly.

'The saints will be in uproar,' Mantell wrote in amusement. He had long subscribed to the view, which was gaining momentum within geological circles, that Moses should be seen as a moral authority rather than be taken literally.

During that summer of 1829, his plans to lead in the field of fossil reptiles were taking shape. By August he was drafting a catalogue, and in September he was ready to open the doors of his home to the public. Great celebrations were organised over two days, and a large party was invited.

But despite the appearance of success and the conviviality of the opening party, all was not well on the domestic front. Although for

Mantell these priceless relics were of 'matchless beauty', for his wife their charm had long since gone. Far from being a genial family home, the atmosphere was cold, learned, scholarly, every inch of space filled with the broken remnants of ancient cold-blooded creatures. The basement was occupied by the domestic staff. The ground floor was given over to the medical practice. His hobby was fast taking over the upper rooms, leaving less and less space for their growing family.

To make matters worse, the saurian bones served as an irresistible magnet drawing a continuous stream of visitors to the house. Although the announcement in the local paper stated that Mantell's museum could be seen only by prior appointment in the afternoon of the first and third Tuesdays of the month, this was invariably ignored. 'Worried to death with visitors,' he noted in his diary, 'this notoriety is a curse . . . my regulations are daily infringed upon to the great annoyance of Mrs Mantell.' Even he was forced to admit that some of the visitors were no more than 'idlers', while others were so 'highly delighted' by his finds that they might stay until one or two o'clock in the morning. Despite all this, and no doubt to Mary Mantell's disgust, her husband was still, on occasion, bringing home fossils – by the cartload!

The 'Mantellian Museum' was well previewed. One of the first to visit was Robert Bakewell, the author of *Introduction to Geology*, a book that had sold well since it first appeared in 1813. When Bakewell saw the bones of reptiles 'of enormous magnitude' occupying an entire side of the museum, he was clearly impressed. Mantell, he said, would 'ride on the back of his Iguanodon into the Temple of Immortality'. In the *Magazine of Natural History*, Bakewell described three days spent in 'much satisfaction' surveying the interesting objects, many of which were 'unrivalled and unique'. The fossils, he said, had been prepared 'with a degree of science and care that I have noticed in no other museum'. Convinced that 'the labours of Mr Mantell did not . . . receive the attention they justly merited', he went out of his way to promote Mantell's career, and introduced him by letter to a key correspondent in America, Professor Benjamin Silliman, at Yale. Silliman was the first to teach geology in America and had founded the *American Journal of Science and Arts.* He was keen to exchange fossils with a British collector.

On 3 November 1829, with the museum established, Mantell began to draft a remarkable paper, to be called 'The Age of Reptiles'. Now completely enveloped in the ancient world he had uncovered, he began to draw together all the threads of evidence. His aim was not only to show that the Weald was of more than just local interest, but also to shed light on an extraordinary era in the history of the globe. As he worked away in the fading light, with the fragments of giant bones forming grotesque shadows all around him, he tried to find the right words to convey the alien picture he had in his mind: 'Among the numerous interesting facts which the researches of modern geologists have brought to light, there is none more extraordinary and imposing than the discovery that there was a period when the *earth was peopled by quadrupeds of a most appalling magnitude* and that these reptiles were the *Lords of Creation*, before the existence of the human race!' (his italics).

Mantell could not know that the dust from the bones surrounding him, settling imperceptibly on the shoulders of his jacket, was well over one hundred million years old. In the late 1820s there was no way of identifying the exact age of the rocks; radioactive decay analysis was not to be discovered until well into the next century. The age of the earth was still a question of belief rather than proof, although by now geologists were no longer writing in terms of the short time-scale outlined by Archbishop Ussher. As Thomas Hawkins pointed out, 'we have lost the ante-diluvial measure of Time, about which Ussher led us into so many lamentable follies . . .' But although Mantell had no way of proving how long ago his giant reptiles thrived, he could define the 'Age of Reptiles' in terms of its *position* in the sequence of rocks. 'The geological period when the existence of reptiles commenced must, according to our present state of knowledge, be placed immediately after the formation of the coal measures [the Carboniferous period].'

It is now known that the Age of Reptiles, called the *Mesozoic era*, which lasted from approximately 245 million to 65 million years ago, began well after the huge coal deposits – the accumulations of millions of years – were forced up and exposed. As these were eroded, carbon dioxide – nowadays known as the greenhouse gas – escaped into the atmosphere, creating the much warmer conditions in which reptiles and dinosaurs could flourish. The Mesozoic era is now divided in three periods, Triassic, Jurassic and Cretaceous. From his knowledge of the rocks of the Secondary period, Mantell could anticipate all three. Some fragmentary remains of lizards and crocodiles, he pointed out, had been found above the coal measures in some of the oldest rocks in the Secondary series: the bituminous slate of Thuringia on the Continent and the New Red Sandstone in England. The New Red Sandstone corresponds to the *Triassic* period of 245–208 million years ago, the earliest part of the Mesozoic era. In 1829, very little was known about the creatures in these ancient rocks.

'It is not till we arrive at the rock termed "Lias" that the remains of reptiles occur in any considerable quantity,' Mantell wrote. With the Lias, he was glimpsing back to 208–145 million years ago, to the dawn of the *Jurassic* age. 'At that period the earth must have teemed with oviparous [egg-laying] quadrupeds,' he marvelled, 'and those which inhabited the sea appear to have been equally numerous with those of the land and rivers.' From very little evidence, he was able to recognise a whole ecosystem. He imagined an ancient time where giant reptiles ruled the world.

Mantell went on to summarise which creatures had been found in the different layers of Secondary rock. In doing so, he implied that there was an order in which these ancient beasts had appeared on the earth, although he did not speculate on why this might be. The remains of two extinct sea lizards, he observed, *Ichthyosaurus* and *Plesiosaurus*, could be found in abundance in the lower beds of the Lias of the early Jurassic period, and inhabited the ancient oceans alongside other creatures such as ancient crocodiles, salamanders, tortoises and turtles. At around the same time, 'several species of the Pterodactylus, or flying reptile, first make their appearance'. Because of the abundance of marine shells such as ammonites, belemnites and nautilites in these lower Jurassic strata, he regarded the rock as having been deposited by an ocean. 'The only apparent exception are the Stonesfield beds in Oxfordshire . . . where we first meet with the remains of the gigantic Megalosaurus.' He envisaged that *Megalosaurus* was contemporary with the sea lizards, and lived on land with ancient crocodiles, insects and even perhaps a few terrestrial mammals, since the tiny jaws of an animal allied to the opossum had also been found. 'The occurrence of terrestrial mammalia in beds of this ancient epoch has not been satisfactorily explained,' he noted.

He then showed that the Age of Reptiles extended through to younger rocks of the Weald in Sussex. These were formed in the third period of the Mesozoic era, now known as the *Cretaceous*, of 145–65 million years ago. The quarries in the Tilgate Forest were then part of a huge freshwater estuary stretching across southern England. Mantell explained that in this epoch strictly marine lizards, such as *Ichthyosaurus*, were rare; but turtles, *Megalosaurus*, one or more species of *Plesiosaurus*, several species of crocodiles and probably pterodactyls could be found. 'At this epoch we also have an enormous herbivorous reptile, the Iguanodon.' He observed that the plants associated with *Iguanodon* are all tropical, and he also hinted that the teeth and bones of other gigantic reptiles had been found which could not yet be identified.

In more recent deposits still, such as the ocean of chalk above the sands and clays in which the Iguanodon was found, he wrote, 'the reptiles are less numerous, and the Megalosaurus, Iguanodon, and other herbivorous genera disappear altogether'. No traces of their existence can be found after this. 'With the chalk, the Age of Reptiles may be said to terminate. The greater part of these genera appear to have become extinct during the changes which took place on the surface of the earth at that period. The crocodiles and turtles &c alone survived, a new order of things commenced and in the Tertiary formation which succeeded we perceive an approach to the modern condition of the earth.'

Gideon Mantell's paper, written in the autumn of 1829, was the first to set out detailed evidence for an Age of Reptiles, which preceded the Age of Mammals, highlighting the order in which beasts appeared on the earth. Although Cuvier had acknowledged that reptiles came before mammals in the development of vertebrates, he did not fully expand this into a coherent argument. It is possible that this was because he did not want to give more ammunition to his opponents, the early evolutionists such as Lamarck and Geoffroy Saint-Hilaire.

If there really had been a reptilian epoch before mammals arrived on the earth, this would lend weight to the evolutionists' view that there had been some kind of progression of life, from primitive to more advanced forms. In his works, Cuvier avoided any discussion of the apparent progression of animal life in the fossil record, which his own brilliant anatomical techniques had helped to bring to light. Although Mantell did not support the evolutionists and could not explain the chronology, his intimate knowledge of the strata and the strange giant creatures found within them made the Age of Reptiles very vivid to him. He summarised the rocks in Europe in which reptiles had been found, hinted that there was an order in which they appeared, established that the earth at one stage was dominated by these creatures, and showed that there was a period when they all disappeared from the rock.

When his ideas were published in the *Edinburgh New Philosophical Journal* in 1831, Mantell hoped that his account would generate interest and financial support. But within one small circle there were some shocked responses. A highly respected scholar, the Reverend William Kirby, scoffed at 'Mr Mantell's Hypothesis of an "Age of Reptiles"'. The notion 'that the Saurians were the mighty masters, as well as monsters of the primaeval animal kingdom and the Lords of Creation before the existence of the human race,' he urged, 'cannot be reconciled with the account of Creation of animals as given in Genesis.'

For the Reverend Kirby, Mantell's hypothesis was flawed on several counts. Firstly, geological evidence already showed that the *Megalosaurus* had been found with the fossil of a mammalian opossum, 'a fact that militates strongly against an insulated Saurian reign,' he reasoned. Kirby also maintained that the remains of the mighty lizards were not just found in ancient deposits, but also in more recent strata. 'The internment of these animals was therefore various,' he said, 'and the facts uncertain . . . so that no satisfactory hypothesis can be built upon them.' Secondly, 'as regards the numbers of these animals, which Mr Mantell thinks prove their prevalence . . . it surely cannot be safely affirmed that for one individual found in a fossil state, thousands must have been devoured or decomposed. These mighty monsters were more likely to devour than to be devoured; even the herbivorous ones, such as the vast Iguanodon, supposed to be sometimes one hundred feet long, would have puzzled the crocodiles and other carnivorous ones to overpower and despatch them.'

Kirby's greatest objection was the challenge to God's purpose that Mantell's hypothesis suggested: 'Who can think that a Being of unbounded power, wisdom and goodness should create a world merely for the habitation of a race of monsters, without a single rational being in it to glorify and serve him! The supposition that these animals were a separate Creation, independent of Man and occupying his eminent station . . . long before Man was brought into existence . . . dislocates the entire system recorded with so much majestic brevity in the first chapter of Genesis.'

Another clergyman, too, wrote to Mantell with similar objections. Man had been resurrected by God and that was why we could find no evidence of Man at this time: there was no such thing as an Age of Reptiles. 'From our knowledge of the works of the Deity,' he claimed, 'it seems improbable that so stupendous a fabric as the Globe of our Earth is, could remain for any period of time solely occupied by a Race of Reptiles.'

However, within the geological fraternity Mantell's ideas were well received. Soon after this, Robert Bakewell described him as 'I think beyond doubt the most scientific anatomist in England . . . a British Cuvier'. Other geologists increasingly sought his advice on unknown fossils: Roderick Murchison, for instance, presented him with a fossil fox, hoping he would identify the species. Charles Lyell, too, on another occasion, wrote of Gideon Mantell's 'genius'. Yet his ideas were still failing to reach a wider audience. Although he published his account of the Age of Reptiles in several other sources, including the *Scientific*

Annual, Silliman's *American Journal of Science* and the *Sussex Weekly Advertiser*, to his dismay there was little further response.

Ironically, he soon found that his writing actually hindered his research. The quarries at Whiteman's Green in the Weald now attracted rival geologists who were quite prepared to outbid him on fossils. The most zealous amateur collector was a Mr Robert Trotter who lived close to Whiteman's Green. Mantell despaired. 'Drove to Cuckfield and endeavoured to obtain some fossils from the quarrymen who have been employed by me so many years, and the ungrateful scoundrels refused to let me have one, having found a customer on the spot. Here is an end to all my hopes of discovering the jaw of the Iguanodon!' His plans still centred on his research on fossil reptiles. If he could not raise money through writing, was there some other way of obtaining funds?

In June 1830, George IV died suddenly. When William IV ascended the throne, in a break with tradition he announced that he and Queen Adelaide would reside at the Pavilion, their palace in Brighton, Sussex. Two months later, the King and Queen made their first visit to Brighton. According to the *Sussex Gazette*: 'an almost countless throng, sixty thousand people, gathered to welcome their Majesties on that joyous occasion'. The small seaside town, barely more than a fishing village ('Brighthelmstone' a century earlier), gradually became a magnet for the fashionable and the wealthy. It was not uncommon to see the King and Queen taking an airing in an open carriage, the carriages of the nobility lining the promenade by the Old Ship Hotel and the cobbled streets around the lavish oriental façade of the Pavilion. For a few months each year, Brighton became the centre of the Season, with Society balls, dinners, parties and the races. With the court less than ten miles from Lewes and the nobility within his sight, Gideon Mantell's long-held hopes of patronage for his research were rekindled.

In October of the same year, the King and Queen were to pay a visit to Lewes. For Mantell this was the perfect opportunity to try to interest the King in this remarkable new science. He sent copies of his books to His Majesty with a message, and to his great delight he received word that the royal party wished to visit his museum in Lewes. To

Detail of Gideon Mantell during the visit of
King William IV to Lewes, 22 October 1830.

prepare themselves for this honour at Castle Place, the maids were
instructed to spring clean everywhere, the silver was polished, the tables
were covered with the best fossil specimens, the bones and casts were
neatly labelled, and the children were bustled into the kitchen. With a
sense of great expectation, Mr and Mrs Mantell awaited the royal
couple's visit.

They could hear the excitement, in the distance: 'the greatest bustle
prevailed . . . the procession passed at a foot's pace the whole length of
the town amidst the thundering of cannon and the ringing of bells . . .
the band struck up a national air'. Inside the house it was quiet, still,
everything perfect for the royal visit. But there was no sound of a
carriage pulling up outside; no noisy crowds approaching; no hurried
footsteps on the stone porch.

Eventually a messenger knocked at the door. The ceremonies in the town had taken so much time, he explained, that it was now too late – 'their Majesties would honor Dr Mantell with another opportunity'. Mantell left at once, struggling through the crowd to the mansion in Friar's Walk where, he knew, the King and Queen would be attending a banquet. Afterwards, the most important local dignitaries and church-men were presented to the King, who was sitting on a magnificent carved chair, surrounded by nobility.

Gideon Mantell was summoned by Sir John Shelley. He stepped forward and knelt on one knee.

'Sire, Your Majesty's gracious condescension in deigning on a former occasion to accept of my Geological Works, induces me to hope that Your Majesty will permit me to lay at your feet, the History of this, my native town.' He politely gave him a short history of Lewes that he had written, whereupon His Majesty, almost interrupting him, said, 'Certainly, certainly, much obliged, much obliged.' Then, turning to Lord Howe, the gentleman-in-waiting, he added impatiently, 'Take them, take them.' Mantell was bustled on, to make way for the next dignitary.

The audience had been over in an instant. The opportunity was lost. All their preparations were for nothing. It was just one more hurtful disappointment, adding to the weight of so many that had gone before. The frustration of never making much headway with his scientific research was becoming almost intolerable. 'My brief existence is running down . . . my time fleeting away and sadly, sadly unimproved, a dull round of visiting; scenes of misery constantly before me which affect me as acutely as ever: scribbling verses, letters, geological scraps and a hundred other nothingnesses.' For all his gargantuan efforts, he still had no patron, and no money from his geological works.

In Sussex in the early 1830s, a serious cholera epidemic broke out. To allay the fears and superstitions which surrounded the disease, Gideon Mantell drafted a guide, *Short and Plain rules for the Prevention and Cure of the Cholera Morbus*. In it he showed how to deal with cholera, and 'called the attention of the rich to the condition of the lower classes, among whom if the disease should occur it is impossible the former should escape'. His pamphlet was applauded by the *Lancet*: 'Mr Mantell . . . has expressed himself so clearly and judiciously that his arguments are calculated to produce the best effects upon the minds of unprofessional readers.'

In the summer of 1832, while the cholera epidemic was at its height, Mantell heard news of an unexpected find at the quarries in the Tilgate Forest. Workers had been blasting the hardest rock when they noticed fragments of petrified bone. Their explosion had caused so much damage, they realised, no amateur was likely to buy the material. Consequently, instead of informing their new customer Mr Trotter, a letter was despatched to Mantell, the only buyer with sufficient expertise to tackle such material.

'On repairing to the quarry, the considerable number of pieces into which the block was broken, the extreme hardness of the stone and the unpromising appearance of the fragments of bone that were visible, seemed to render the attempt to dissect it alike hopeless and unprofitable,' Mantell wrote. He had hoped for part of the jaw of the *Iguanodon*: 'if I find but that before I die I shall be content'. But the deeply embedded fossils did not appear to match the bones or teeth of an *Iguanodon*. Patiently he set to work, gathering more than fifty scattered fragments together. Although aware that this would not please his wife, he made arrangements for the large blocks of stone to be brought by cart to his home in Castle Place, Lewes.

For several weeks he worked at night, isolating the bone from the rock and trying to fit the pieces together until he had a huge slab over four and a half feet long. Gradually he was able to reveal part of a spine, with several vertebrae, ribs and sternum. There was also something that he had not seen before: to the left of the vertebral column were large bony attachments which served no apparent purpose and were quite unlike bones in the *Megalosaurus* or *Iguanodon*. There were at least ten of these strange bones, reaching seventeen inches in length and up to seven inches wide at the base.

Intrigued, he devoted his 'little leisure . . . to the labor of chiselling out the magnificent specimen'. The vertebrae and ribs bore most resemblance to the corresponding parts of a crocodile. The sternum was more similar to that of lizards. Most puzzling were the flat, pointed bones – they corresponded neither to the shell of a giant turtle nor to the protective scutes, or bony plates, of the giant armadillo. With great insight, he came up with a radical interpretation. These bones, he thought, must have run down the length of the spine, like primitive body armour.

He also found bizarre 'dermal bones', like very thick scales, which he thought formed plates in the skin. Although there was no analogy for such a creature, he realised that this was yet another type of giant reptile, new to science, equipped with heavy, bony armour for defence. 'I have made a grand discovery,' he wrote proudly to his friends. It was, in fact, the first of the armoured dinosaurs, now known as ankylosaurs. Mantell named his new reptile the '*Hylaeosaurus*', or forest lizard.

Such was his standing, now, in scientific circles that he was soon invited to describe his find at a meeting of the Geological Society. He sent his celebrated new fossil to Somerset House in London, together with many giant bones and a large painting of the hind limb of the *Iguanodon*. 'A very full meeting,' Mantell observed, 'all my friends were there, my kind friend Mr Bakewell, though infirm, had even ventured out . . . All passed off very well and at the conclusion I had the painting let down, which very much gratified the greater part of the audience.'

Absorbed in his presentation, Mantell scarcely paid any attention to a relative newcomer in the field of fossil reptiles, sitting quietly in the audience: Richard Owen, the young assistant at the Hunterian Museum. Owen had been 'devoted' to preparing John Hunter's collection; the weeks had turned swiftly into months as he planned the many lengthy catalogues. In physiology alone, he and William Clift were drafting separate volumes on the organs of digestion, on circulation, on the respiratory and urinary systems, on the nervous system, and on the organs of sense and reproduction. There was also a pathological series to illustrate disease processes, which Clift personally supervised, not to mention a collection of 'monsters and malformed parts' that his son,

William Home Clift, was describing. In addition to this, for each series they had numerous species to classify, from both fossil and living animals. Owen seemed to blossom under this labour. With each dissection, and as he absorbed the works and thoughts of the famous anatomist, the mantle of John Hunter fell more surely about his shoulders.

At the Geological Society presentation, as Owen's eyes feasted on the extraordinary fossils that Gideon Mantell displayed, he might well have been considering how he could acquire such riches for the museum at the Royal College. He took in the whole scene: Mantell's brilliant interpretations of the fossils, the appreciative remarks of Mr Lyell and Mr Fitton addressing the Society – though, in 'too partial a strain in Mantell's favour' – and the resounding applause of the learned gentlemen when he had finished.

Mantell had reason to feel content. He had found and identified two of the three giant land reptiles known to science, *Iguanodon* and *Hylaeosaurus*, and defined the Age of Reptiles with more clarity than anyone else. It had taken him years to arrive at this point, but now he was acknowledged and respected by many as the leader in the field. The fact that the young Richard Owen was fast acquiring the skills in comparative anatomy that were crucial to the subject was lost on him. Mantell had no reason to suspect that his unlined face and charming manner concealed an increasingly well tutored ambition.

9

Nature, Red in Tooth and Claw

> Man . . .
> Who trusted God was love indeed
> And love Creation's final law –
> Tho' Nature, red in tooth and claw
> With ravine, shrieked against his creed.
>
> Alfred, Lord Tennyson, *In Memoriam*

Richard Owen had run into an obstacle to his plans to marry Mr Clift's daughter. Despite his high hopes of taking Caroline for his bride, months passed and he was still no nearer satisfying his prospective mother-in-law – in particular, her financial requirements. After all, he was clearly not a gentleman of inherited wealth, his earnings were modest, and his prospects even more so while he remained at the Royal College, because the Clifts' only son, William, had been promised that on his father's death he would inherit the one prominent post at the Hunterian Museum, that of Conservator. It was clear to Owen that there was little chance of advancing in his career or his marriage unless he moved elsewhere.

Even with such an incentive, Owen spurned posts that would have offered a faster route to financial independence if it meant sacrificing his consuming interest in anatomy. Two years into his engagement to Caroline, when a position became available at the Birmingham Hospital he shuddered at the prospects of arduous routine medicine with no research. He explained quite frankly to Clift that he could not face 'ten long years' fag and saving of scraps, away from those I love most and the society I take such delight in'. He was in no hurry to give up his intellectual freedom at the College to study anatomy *just* to secure the hand of his bride.

By understanding the anatomy of creatures, assessing their affinities and classifying them, he hoped to make inferences about their place in Nature and how they formed: to bring order to the wild profusion of the animal kingdom. At stake was the baffling puzzle of how life was created. Why was there a succession of 'former creations' in the fossil record?

In the early 1830s, as the geologists produced yet more evidence to support this by showing that an Age of Reptiles seemed to precede the Age of Mammals, the pioneering anatomist Étienne Geoffroy Saint-Hilaire in Paris was still making provocative new claims for his early evolutionary ideas. For several years, he had been suggesting that living creatures were 'descended by an uninterrupted path of generation' from fossil predecessors. He speculated on a possible order of beasts. Among the reptiles, *Ichthyosaurus*, *Plesiosaurus*, *Pterodactylus*, and *Teleosaurus* had 'progressed' in some way to the extinct giant mammals of the Tertiary rocks such as *Megatherium*. Georges Cuvier considered these evolutionary ideas fundamentally flawed. For him, the animal kingdom could be divided into four major 'branches', or groups, each of which was so anatomically distinct that they could not possibly be compared.

Geoffroy pushed further and further into uncharted territory, seeking equivalent parts, or 'homologies', between different classes of animals to prove that they were linked. His thinking was adventurous, flamboyant, sometimes even absurd. He attracted Cuvier's ridicule by suggesting that the carapace or upper shell of insects corresponded to the vertebrae of vertebrates. But there was no evidence to support this speculation, and Cuvier used the opportunity to dismiss him as a mere 'poet'. Undaunted, Geoffroy stalked the intellectual territory, seeking transitions between the different groups of animals to further the case for progressionism.

In February 1830, Geoffroy presented a bold idea to the Academy of Sciences in Paris. There were homologies, he claimed, between some vertebrates such as fish and certain invertebrates known as cephalopods, a class of marine molluscs that includes cuttlefish, octopus, squid and the fossil ammonites and belemnites. His hypothesis immediately attracted attention. If it was true it implied that evolution between two of Cuvier's four branches of the animal kingdom – the 'higher' vertebrates and the 'lower' molluscs – *was* possible. Cuvier was so appalled that he used his political weight to block the examination of Geoffroy's ideas. Since all civil servants were obliged to support religious beliefs, Cuvier's attacks were potentially dangerous for Geoffroy, and their feud erupted into an acrimonious public debate which was widely reported that spring. Fears that such radical philosophies could help to incite rebellion resurfaced during the July Revolution, in which Charles X fled Paris.

The early evolutionary ideas developed by Geoffroy in France continued to receive a hostile press elsewhere. According to the English *Monthly Review* of 1832, evolution was 'the most stupid and ridiculous' idea to have been hatched by 'the heated fancy of man'. By undermining the authority of the Bible, the evolutionists appeared to challenge the very social and moral foundations of society. The debate was inflamed in England by a background of social upheaval and threatened authority. There were widespread agricultural riots, and the new industrial centres saw a wave of violent protests. The immense wealth of the aristocracy was deeply resented, and the Whigs came to power amid clamour for reform and fears of an English revolution. The 1832 Reform Bill redistributed power, increasing parliamentary seats to industrial towns and enabling many more householders to have the vote.

At the Royal College, following in Hunter's and Cuvier's footsteps, Richard Owen was keen to use the subject of anatomy to counter the French progressionist ideas. His chance came in 1831, in the form of a rare sea creature sent to the Hunterian Museum from Polynesia: the pearly nautilus. This exquisite creature with a spiral chambered shell and pearly interior belonged to the class of invertebrates that Geoffroy had claimed showed 'homologies' to vertebrates, the cephalopods. Peering into the bottle holding the specimen, Owen could see an opportunity. This beautiful creature was a rare gem with which he could dazzle the scientific authorities.

In a sixty-page study, he launched his attack on his radical rivals. He highlighted the animal's uniqueness. The pearly nautilus was a creature as 'rich in the variety of parts as it is peculiar in its mode of arrangement'. Far from Nature forming an unbroken series that paved the way for evolutionary claims, Owen maintained the reverse: the mollusc's anatomy differed too much from that of the vertebrates for there to be any connection. Owen was hoping to seek Cuvier's favourable opinion on his paper, but events took an unexpected turn.

Georges Cuvier had also been looking for an opportunity to tackle his ungodly opponents. On 8 May 1832, when invited to give a public lecture at the Collège de France, he condemned the 'pantheism' of Geoffroy Saint-Hilaire and disclaimed his rival's 'useless scientific theories'. Carried away in the heat of the moment, he made an impassioned speech on the Divine Intelligence within natural science.

While his talk made a tremendous impact and his audience was reported to be 'overcome with emotion', the ageing Cuvier, too, was overwhelmed – by the effort. That evening he experienced the mysterious symptoms of slight paralysis – probably from a stroke – which were to take his life six days later. The sudden death of so eminent a man, a peer of France, grand officier of the Legion of Honour, the great 'Baron' of natural science, was a loss deeply felt in scientific circles. It created an immediate vacuum, both on the Continent and in England. The throne from which the 'Napoleon of Intelligence' had dominated the thinking of a generation in the fledgling sciences of geology and anatomy was waiting to be filled. Who would be the next Cuvier?

Richard Owen, coveting the title for himself, used his study of the nautilus to launch himself on to the scientific stage. 'Since the decease of the lamented Cuvier, there is no one whose opinion on this work I look for with more anxiety than your own,' he told the Reverend Buckland. Buckland replied that he was 'highly gratified' by the 'masterly insights' of his 'admirable memoir'. Sir Anthony Carlisle, a former College President and member of the Royal College Council, also showered praises on their young protégé. 'It is an excellent specimen of Hunterian–Cuvierian Natural History,' he wrote, 'but as I foresaw, your pearls are thrown before swine. If the English hog trough should be cleared out in our time, there is a gleam of hope for science among a small few, but you must not feel disappointed by the general neglect of your researches.'

But to Owen's dismay, the *Lancet* predicted that another scientist would be 'the next Cuvier': the anatomist Professor Robert Grant of the University of London. According to Thomas Wakley, the journal's editor, 'Grant displays a perfect *mastery* over his vast subject' (his italics). Wakley could not commend highly enough 'the integrity, capacity and vigour of his mind'. Indeed, there was no other like him 'in the entire British dominions'. And to make matters worse, Professor Grant wholeheartedly embraced the evolutionary thinking of the French. According to science historian Adrian Desmond, the new University College in London – in contrast to Oxbridge – was the 'Godless' college, open to students of any faith, where Grant was free to debate Geoffroy's ideas. Like Geoffroy, Grant envisaged that as the primitive earth slowly cooled, the resulting changes in climate had created new habitats which enabled life to progress into the great diversity of living forms. From the idea of evolution beginning with the simple marine sponges – a clue, surely, to primitive forms of life – to the heretical notion that Man might have 'progressed' in some way from chimpanzees, for Grant no line of enquiry was sacred. In his packed lecture theatre, the radical ideas from Paris, for so long held at bay by Anglican dons, were unleashed into the study of biology.

To counter this challenge, Owen needed more specimens for dissection, to prove the errors in the evolutionary thinking of Geoffroy and Grant. It was soon apparent that there was no better place to acquire them than at the Zoological Society of London. In the 1830s this was something of an elite gentlemen's club where dukes and duchesses could select rare creatures from Britain's burgeoning Empire to adorn the parkland of their country estates. Owing to the lack of experience of exotic animal husbandry, deaths at the zoo were only too common. In the space of a fortnight, Caroline Clift recorded in her journal: 'Poor George, the lion, dead . . . The sloth bear found dead with his two companions doing their best to eat him . . . one of the dingkos escaped.' Soon after this, the wild ass 'was gored so horribly by a Wapiti stag the keeper was obliged to put it out of its misery'.

When Owen joined the Zoological Society in 1830, to his concern he found his rival, Professor Robert Grant, already well established there. Grant, like Owen, hoped to make his name from pioneering studies on the creatures whose anatomy had so far eluded the gentlemen of science. Such was his standing, Grant was chosen to deliver the first lecture series on anatomy to the learned members. But it wasn't long before the young Owen manoeuvred himself on to the Society Council. Cultivating his aristocratic contacts at the Royal College, he searched for opportunities to increase his power within the Zoological Society. Working with the Secretary, he helped to arrange evening meetings and the publication of papers, including his own. Instinctively predatory, he was looking for a way to dispose of Grant and the evolutionary ideas he promoted.

But Owen's intellectual ambitions still cut no ice with his future mother-in-law, who wanted real wealth for her daughter. He was now five years into his engagement to Caroline, and Mrs Clift was concerned that despite his protestations of love for her he had not distinguished himself with any serious effort to increase his income. In 1832, Owen wrote to Caroline entreating her help 'in endeavouring to abridge the term that opposes itself to our union'. He was evidently not lacking in self-assurance, for he continued: 'At present our ruling Goths are blind to what everyone else sees, which, to speak very modestly, is my merit.' More 'enlightened' members of the College, he reassured Caroline, were thinking of creating a permanent professorship for him, perhaps in three years, which would more than double his salary, 'and then with what happiness should I clasp my dear Caroline . . . Now, said I to myself, what is to hinder my dear Cary and me from quietly enjoying ourselves in a more humble way in the meanwhile . . . Now will you write to her [Mrs Clift] or speak to her?' As Owen explained, he could not work well until 'I can "calm this troubled breast" and call you indeed my own'. Mrs Clift, however, remained blind to his charms.

By chance, later that year, a tragedy occurred that dramatically changed Owen's fortunes, through no effort on his part. On 11 September 1832, the young William Home Clift was returning to the College one evening by hansom cab. As the cab swung out of Fleet Street into the narrow Chancery Lane, the driver misjudged the turn and tipped the cab over. Although a minor accident, it was a major catastrophe for Clift. He was flung violently on to his head and taken unconscious to St Bartholomew's Hospital. Here, he was brought to the one man who might help: Richard Owen.

But there was little Owen could do. Clift had a fracture at the base of his skull and there was no treatment to clear up the infection that set in. William Clift senior was enjoying a rare break out of town and had no idea of his only son's fate. Since he was travelling it took several days to track him down. The injury to William Home Clift's brain destroyed him slowly and surely. After the young man had lingered a few days, his father arrived home only to find him close to death. The loss was a 'great grief' to Mr Clift.

The vacant place of only son was there to be filled, and as time went by, Owen came neatly to fill it and was embraced and encouraged like a son by Mr Clift. His career, for so long at a standstill, began to take off. In a few months his pay increased to £300 a year, a level almost comparable to the pay of the Curator. In 1833, the first volumes of the Hunterian catalogue were published, and were praised as conferring honour on the College. Powerful supporters, such as the former President, Sir Anthony Carlisle, wrote to Owen to express their delight. 'I will use my best endeavours to promote your welfare in the College and out of it,' Carlisle promised.

Owen continued to use the powerful tool of anatomy to attack the early evolutionists. His next target became a group of animals known as 'monotremes', which includes the duck-billed platypus and the spiny anteater. Geoffroy in Paris, seeking support for his evolutionary ideas, claimed that these were *transitional* animals, part reptile in that they laid eggs and part mammal in that they were warm-blooded. In a brilliant

The Hunterian Museum.

series of experiments, Owen showed that they were not transitional creatures, but primitive *mammals*.

The debate came to centre on tiny glands in the platypus that secrete a milky substance through the skin. Geoffroy maintained that these were scent ducts. Owen dissected five female platypuses sent from the Australian colonies and proved that the size of the anomalous gland was related to the ovarian cycle: the glands were most enlarged when eggs had just been shed from the ovary. The mysterious glands were therefore mammary glands, a characteristic of a mammal. Within the Zoological Society and the Royal College, Owen's ingenious study was seen as a triumph. It symbolised Britain's imperial prowess and the growing wealth of natural history collections returned from the colonies. Better still, this was the first time the British had dethroned the French in the field of anatomy.

Owen's work was beginning to attract the interest of powerful, like-minded allies, notably Professor William Buckland, who wholeheartedly approved of the younger man's unshakeable faith that the study of anatomy would reveal the works of God. By now Buckland was at the height of his powers and held in such high regard in both the scientific and religious establishments that he was one of eight distinguished thinkers appointed by the President of the Royal Society to write the *Bridgewater Treatises*. The Right Honourable Reverend Francis, Earl of Bridgewater, had been so disturbed by the advance of more secular ideas in science that he had left £8,000 in his will for the Royal Society to 'appoint persons to write, print and publish one thousand copies of a work on the Power, Wisdom and Goodness of God as manifested in the Creation; illustrating such work by all reasonable arguments'.

Aware of the importance of the task, the Reverend Buckland worked for six years on the *Treatises*, ably supported by his wife, who 'sat up night after night, for weeks and months consecutively, writing to Buckland's dictation, and this often till the sun's rays shining through the shutters at early morn warned the husband to cease from thinking, the wife to rest her weary hand'. Surrounded by his geological trophies, in the drawing-room in which a person 'might range a whole day' and still

Sketch of William Buckland on a geological tour. The rocks at his feet are labelled: 'Specimen no. 1, scratched by a glacier thirty-three thousand three hundred and thirty-three years before Creation; no. 2, scratched by a cart wheel on Waterloo Bridge the day before yesterday; the whole picture being scratched by T. Sopwith.'

find something new, his table made entirely from coprolites, the cabinets so crammed 'they had not been invaded by the dust-cloth for the last five years', he was determined to reconcile the apparent conflicts between the new science and religion. For him, geology was nothing less than 'the unfolding records of the operations of the Almighty Author of the Universe, written by the finger of God himself, upon the foundations of the everlasting hills'.

However, Buckland acknowledged the growing number of paradoxes. Geologists had failed to find evidence for a Flood; still less was there scientific evidence for the Creation story, as outlined in Genesis. There were many extinct species, in rock strata which must have been 'deposited slowly and gradually during very long periods of time and at widely distant intervals'. He tackled these problems by questioning the meaning of the Bible. Verse by verse, he scrutinised Genesis, arriving at ever more convoluted interpretations.

'Nowhere is it affirmed that God created the heaven and the earth in the first day,' he asserted, 'but in the beginning; this "beginning" may be an epoch of unmeasured distance . . . during which all the physical operations disclosed by geology were going on . . . Millions upon millions of years may have occupied the indefinite interval between the beginning in which God created heaven and earth, and the evening or the commencement of the first day of the Mosaic narrative.' Then he redefined the term 'Creation': 'this by no means necessarily implies creation out of nothing, it may be . . . a new arrangement of materials that existed before'. The creation of the sun, the moon and the stars on the fourth day, he said, should be seen merely as a rearrangement of conditions so that the stars became visible to Man.

But even as he laboured to breech the ever-widening gap, new evidence kept emerging in support of the unholy progressionists. By chance, Buckland had mentioned to Roderick Murchison, the then Secretary of the Geological Society, that it might be possible to unravel the sequence of rocks below the Secondary strata at a site on the Welsh borders. Murchison held Buckland's knowledge of strata in such high regard that he went to investigate. Although the Secondary strata and their characteristic fossils were becoming well established and had been described by Mantell in 'The Age of Reptiles', the Transition series of rocks below were relatively unknown. At the Welsh borderland, Murchison found a site where he could trace the layers down, deeper into the earth's crust. Starting well below the coal measures (the Carboniferous period) in rock known as 'Old Red Sandstone', he went down into the Transition series below.

These ancient rocks revealed an unfamiliar landscape. There was no sign of terrestrial life, land plants or vertebrates. The fossils were marine, and they told of strange forms of existence unlike any living species. There were odd sea creatures called 'trilobites', up to six inches long, with a segmented skeleton and large multilensed eyes. These had first been systematically described in 1822 by Alexandre Brongniart in Paris; Murchison now placed them at the heart of an entire ecosystem. Trilobites shared the primeval seas with other invertebrates: 'crinoids', which resembled sea lilies and had stalks attached to the sea floor, and 'echinoids', which were similar to starfish, sea urchins and sea cucumbers. Murchison called the rocks 'Silurian' after the Silures, a clan which had thrived in the Welsh borderlands two thousand years previously. In the Silurian rocks it was possible to glimpse a period in the earth's history when simple forms of life, such as marine invertebrates, once dominated.

When Murchison described his finds to the Geological Society, he was hinting at a disturbing idea. Even though there were still many gaps in the fossil record, the progression of life, for so long obscured by folding and faulting in layer upon layer of Transition rock, took on an awesome clarity. The Primary rocks, the oldest of all, were known to have no fossils. Above these, in the Transition series, Murchison's evidence suggested there were primitive life forms – trilobites, invertebrates and plants – in former oceans. Much later, there was an Age of Reptiles entombed in the Secondary rocks, with lizards and salamanders in the lower rocks and then gigantic forms of reptilian life. Above this, in the Tertiary rocks came the Age of Mammals, in which Man himself finally appeared, the pinnacle of Creation.

This progression gave powerful support to the arguments of the early evolutionists like Geoffroy Saint-Hilaire and Grant. Yet, even in the face of the evidence for this order, Lyell still maintained that progression could be an illusory idea. He believed that the fossilisation of vertebrates within Secondary and Tertiary strata was very unreliable and should not be taken as evidence. With vertebrates out of the picture, there was no progression. Anomalies had to be explained, too. How could mammal jaws such as those of the Stonesfield opossum find their way into the Age of Reptiles? As for the older Primary rocks, these, Lyell suggested, only lacked fossils because they had been destroyed as the rocks were formed.

William Buckland and other Anglican leaders of science tackled the ungodly progressionists differently. For them the fossil record provided evidence of a series of Divine Creations as the world was made ever more perfect for Man. Buckland considered that the superb design of the giant *Megalosaurus* or *Iguanodon proved* the existence of a skilled Creator. This 'argument from design' had been cited by classicists in one form or another for centuries: the incredible variety and complexity of living forms, seen throughout the animal kingdom, must reflect the presence of an intelligent Designer.

Buckland used the structure of the *Iguanodon* teeth to illustrate his argument. 'We cannot view such examples of mechanical contrivance united with so much economy of expenditure . . . without feeling a profound conviction that all this . . . has resulted from Design and high Intelligence.' The glorious and complicated design of these ancient beasts argued against the idea that life had evolved from primitive to complex forms. Even the humble trilobites, 'buried for incalculable ages in the early strata of the Transition', had eyes of extraordinary complexity, some species having at least four hundred lenses fixed on the surface of the cornea. 'This is utterly inexplicable,' he declared, without reference 'to the same Intelligent Creative power',

In the *Bridgewater Treatises* Buckland was concerned to explain why the Creator chose to fill the primitive world with evil carnivorous beasts. His very own evidence showed that the reptilian carnivores were furnished with 'organs for the purpose of capturing and killing their prey, instruments formed expressly for destruction'; Nature was hideously red in tooth and claw. He accepted that this was 'inconsistent with a Creation founded in Benevolence and tending to produce the greatest amount of enjoyment to the greatest number of individuals'. Nonetheless, he sought to reconcile even this with God's wisdom. He reasoned:

> It has pleased the Creator to give to every creature upon the earth a dispensation of kindness to make the end of life to each individual as easy as possible. The most easy death is proverbially that which is the least expected . . . by sudden destruction and rapid succession the feeble and disabled are speedily relieved from suffering and the world is at all times crowded with sentient and happy beings . . . the momentary pain of sudden and unexpected death is an evil infinitely small in comparison with the enjoyments of which it is the terminations.

In the Reverend Buckland's reassuring interpretation of the 'unending carnivorous warfare' of the ancient world, the gigantic *Megalosaurus* became, paradoxically, God's agent for reducing the total sum of animal suffering.

In Georgian England his book was welcomed, selling out within weeks. 'It will astonish and delight all lovers of science,' enthused the *Quarterly Review*. With his 'commanding eloquence' Buckland had 'swelled the chorus in which all creation hymns His praise' and was 'witness to His unlimited Power, Wisdom and Benevolence'. The *Edinburgh Review* commented that it was 'calculated to inspire the most affectionate veneration for that Great Being'. For Buckland, it was a personal triumph. In scientific circles he was considered to have successfully defined the ground rules within which geological research could be conducted.

But even in the mid-1830s, there was still no shortage of biblical literalists united in a chorus of disapproval. In *Blackwood's Edinburgh Magazine* one reviewer wrote scathingly of 'a clergyman giddily giving a date and origin to the world wholly contradictory to that which is expressly given in the Bible'. For literalists like George Bugg in *Scriptural Geology*, to contradict Genesis was nothing less than flagrant sin. He considered the scientific view so preposterous that he could not even accept that 'races' of carnivorous animals had once thrived before Adam's downfall.

'Animals were *not created carnivorous*,' stormed Bugg (his italics). 'I hold this to be a most indisputable principle. If animals were created carnivorous, "death", even violent death, must have been common in Creation from the very beginning. But the Scripture represents Death as entering into the world by *Sin*. Had lions, and tigers &c been as voracious from the first as they are now . . . Adam himself would not have been safe from destruction by voracious animals.' It was Adam who brought death and suffering into the world. Before this, claimed Bugg, Adam was not even troubled with lice, fleas or parasitic worms! In his view, *Megalosaurus* and other giant beasts were originally herbivorous, and such creatures 'degenerated from their original state into their carnivorous habits'.

Buckland's *Bridgewater Treatise* was serialised in *The Gentleman's Magazine*, and the resulting publicity prompted even more remarkable speculations. According to a Mr Thomas Thompson Esq., Vice-President of the Hull Literary and Philosophical Society, Buckland's giant reptiles could indeed be reconciled with the Bible. 'There is good ground for supposing that the Leviathan of the Scriptures is the same animal as the new fossil Megalosaurus, and the Behemoth was identical with the Iguanodon,' he wrote in the *Magazine of Natural History*. But others disagreed. 'Will it be said that the Megalosaurus existed in the Nile so lately as the time of Isaiah, about 300 years only before the age of Herodotus?' queried one reviewer in the *Edinburgh New Philosophical Journal*. 'With respect to Leviathan, there is no passage in the Bible in which we can gather from the context that it means any animal other than the crocodile of the Nile.'

In taking a path of reconciliation between science and religion, William Buckland had once again struck a raw nerve.

Richard Owen, quietly studying the *Bridgewater Treatises* at his lodgings in Symond's Inn in central London, was impressed. Although he did not accept that all the creatures in Nature were individually designed by a

Divine hand, he did believe that fundamental laws of anatomy lay behind the different forms of life, laws made by God which were established at the point of Creation. He wanted to understand these 'Divine blue-prints', from which he thought all the myriad different forms of life had sprung. He took a particular interest in Buckland's arguments on the giant reptiles.

If the most *advanced* reptiles with complicated anatomical designs had lived *early* in the history of the globe, this argued against the notion of evolution, which implied a progression from simple to advanced forms. Perhaps, as William Buckland had hinted, the prehistoric giant reptiles could be used as a weapon to silence the evolutionists, once and for all. When he had time, he began to gather information on the giant fossil reptiles. He was soon writing papers noting details of their anatomy, such as 'On the Dislocation of the Tail at a certain point Observable in the Skeletons of many Ichthyosauria'. In this, he correctly argued that these lizards possessed a heavy caudal (tail) fin. Little by little, he was searching for a way to dispose of his rivals.

In 1834, Richard Owen was promoted to Professor of Comparative Anatomy at St Bartholomew's Hospital, and was soon appointed a Fellow of the Royal Society for his papers on the duck-billed platypus and marsupial animals. As he had long hoped, when once it was clear that he was making headway in his career, Mrs Clift's opposition to his marriage collapsed. After an eight-year engagement and within a few months of his joining the Royal Society, she relented, and plans were made for the wedding.

He was now provided with a suite of rooms at the Royal College of Surgeons. It was gratifying for Caroline that her fiancé not only had a home, but a most imposing one. Approached from the leafy square in Lincoln's Inn Fields, up wide stone steps with six towering Doric columns, his apartments were reached through the spacious interior of the College, with its echoing marble floors and pillared hallway. 'R.O. took us to his HOUSE, where he regaled us with ices, claret and cakes,' she wrote. 'I was agreeably surprised at the size of the rooms and the comfort of the kitchens.'

The long-awaited event, on 20 July 1835, was a small family affair held at the new St Pancras Church in Euston Square. Owen's wedding day, which was also his birthday, marked a turning-point in his life. Clift, unassuming, well liked, and respected for his years of patient work on John Hunter's collections, was the perfect patron for his ambitious and talented son-in-law. He would never hear an ill word against him and was always willing to promote his interests. Soon, Richard Owen enjoyed the prominent position of deputy to Clift at the Royal College of Surgeons.

Finally, in April 1836, the hopes that he had so long entertained became a reality. A special post was offered to him at the College: Hunterian Professor. This unique appointment required that he give an annual course of lectures in honour of John Hunter and instantly elevated the young deputy to a very senior rank in scientific circles. He was delighted. 'I beg to express to the Council my deep sense of this additional mark of their favourable sentiment towards me,' he replied. His grandson's biography reveals that 'to the last days of his life, Richard Owen constantly referred to the gratification which this appointment gave him'.

Honoured with this great distinction, Owen seized the moment to dispose of Robert Grant, his long-standing rival at the Zoological Society. Hostilities between the radical, 'evolutionary' Robert Grant and Richard Owen had often erupted into petty disputes over who should dissect which rare specimen. Now backed by his father-in-law and a growing band of supporters at the Zoological Society Council, the young Hunterian Professor's despatching of Grant was effortlessly simple, a coup was achieved in a matter of minutes. One evening, at a Council meeting at the Society's museum in Bruton Street, Owen merely vetoed Grant's appointment to the Council, thereby disempowering him in one stroke. For a comparative anatomist to have no access to the best source of animals for dissection was a major setback. Robert Grant woke up too late to the danger, and tried to rally his supporters. The 'malcontents', as Caroline Owen called them, were unsuccessful. For Grant this marked the beginning of a painful downward slide which eventually spiralled out of control.

By preventing Grant from obtaining specimens, Owen adeptly cut off any hopes he might have of advancing his position in this field, and his reputation waned. As Adrian Desmond has shown, Owen isolated him and deprived him of key sources of information, turning the man who was tipped to be the next Cuvier into an unwelcome visitor at the Zoological Gardens. More and more people joined the establishment bandwagon, championed by Owen, opposing Grant's radical views. His support collapsed, and finally his name disappeared from the Zoological Society register altogether.

In time, Grant also found himself struggling for students, and since lecturers were paid per student, his pay fell sharply – sometimes as low as £50 a year. Records from the 1840s show that he tried to borrow money from the university, and when he failed he moved to 'a slum in Camden town, amid harlots and knaves'. The once brilliant Robert Grant did not make a comeback.

Meanwhile, as Owen cultivated powerful aristocrats on the Zoological Society Council, such as Sir Peter Egerton and Lord Braybrooke, he was eventually able so to arrange matters that he had sole access to any dead animals. 'Affairs were settled satisfactorily at the Zoological Council on the question of dissection of animals,' Caroline noted in her diary. 'An order has been entered to the effect that the Hunterian Professor should be allowed to dissect whenever and whatever he liked when Death occurred at the Gardens . . . and that he is to have precedence over any other person.' Owen's victory was complete.

With Grant and any other aspiring rival – out of the way, Owen could now exploit each animal death at the Zoo to advance his own career. By sheer good fortune, that autumn the supply of important specimens to dissect suddenly swelled even more when Charles Darwin, who had just returned from South America on the *Beagle*, donated eighty mammals and over four hundred birds to the Society. Darwin, who was eager to make his way in the London scientific world, was delighted to have his prize fossils interpreted by Owen.

Caroline's diary records a succession of studies, all with a view to understanding God's purpose in creating the myriad living forms. 'Today, Richard cut up the giraffe which died at the Zoological Gardens. Afterwards he went to the Royal Institution to dissect a snake.' Not long afterwards, 'Richard went to Bruton St, the museum of the Zoological Society, to cut up an ostrich.' A few weeks later she noted: 'Poor little chimpanzee dead. R went to see the "opening scene" in Bruton Street.'

Caroline, born and bred in the museum, took the continual passage of dead mammals through her front door in her stride, even showing forbearance when a large rhinoceros was placed in the hallway. The smell of preserving-spirits pervaded their home, a constant reminder of Owen's work penetrating the very air they breathed. Living over the museum, his output was prodigious, and he did not find it necessary to take a break on social occasions. 'When we got home R insisted upon having the legs of a fowl which we had for dinner, to examine the muscles.' Far from seeming gruesome, to Caroline such dedication was all part of his brilliance.

As Owen's power grew within the Zoological Society, the Royal Society and the Royal College, he was very aware of his own merit, and this abundance of self-confidence formed the bedrock of his personality, which to his rivals seemed like some unassailable cliff-face. His razor-sharp mind and raw energy were attuned not just to dissecting animals but to manipulating power within each institution he joined, all the while disarming any suspicions by his sheer youth. It took time for a pattern to emerge, for people to recognise the ruthless streak beneath the charming veneer. His enormous skill, even at this early stage in his career, was to build a power base in each institution he joined, so that his scientific ideas allied to his political astuteness ensured that he was always the man of the moment.

The same pattern was to repeat itself when a new organisation was formed in the 1830s: the British Association for the Advancement of Science (BAAS), created as a rival to the Royal Society in leading and promoting British science. During the 1820s, the deaths of two distinguished Royal Society presidents, Sir Joseph Banks and Sir Humphry Davy, had opened up a vacuum in the management of science. Banks's forty-one-year tenure as President of the Royal Society had brought few reforms; aristocrats and wealthy gentry continued to outnumber genuine scientists as members, and Banks was even unfairly accused of packing the Council with his favourites. In short, the Royal Society was seen as conservative, London-based and elitist in its choice of Fellows, prompting controversy in the press on the 'Decline of Science in England'.

Unlike the Royal Society's, the new BAAS's annual meetings were to be held in a different town each year. The aim was to open up a wider forum for scientific debate, where talented amateurs in the regions could contribute more easily. But despite the admirable intentions, in practice the Anglican leaders of science from Oxbridge and the powerful inner core of Geological Society members rapidly stepped in to steer the new organisation and ensure that it promoted 'God's order and rule'.

Richard Owen was quick to realise that the BAAS had funds to dispose of and that they were looking for talent. At the same time, mindful of Buckland's arguments in the *Bridgewater Treatises*, he was beginning to recognise that the giant fossil reptiles could be a crucial weapon in his crusade against the evolutionists. The fossil reptiles needed to be classified within the animal kingdom; where did they fit in the vast network of Nature? *Had* they evolved from other creatures, or had they been specially created by God? Clearly, the fossil reptiles were to be the next battleground. But he had a new adversary, widely regarded at the time as the leader in the field: Gideon Mantell.

Owen could never easily challenge Mantell's superiority in the field of fossil reptiles within the hallowed walls of the Royal Society. Mantell was nearly fifteen years older, and had been presenting highly regarded papers at Somerset House since 1825. But with his father-in-law, Clift, holding a prominent position within the BAAS, and his ally William Buckland, President of the BAAS in 1832, Owen could see an opportunity. There was a way of disposing of his new rival as he had disposed of Grant, and claiming the intellectual territory as his own.

10

Nil Desperandum

So the cheek may be tinged with a warm sunny smile,
While the cold heart to ruin runs darkly the while.

Thomas Moore, cited in Gideon
Mantell's correspondence, 1836

Gideon Mantell was looking for a way of capitalising on his lead in fossil reptiles in order to advance his scientific career. As he pondered his future, two distinct possibilities began to take shape in his mind. He could keep his country practice in Lewes, where, he felt, 'like a torch I consume myself' in frustration at the lack of time for research. Alternatively, he could move to the fashionable seaside town of Brighton. Surrounded by aristocrats at the court, Mantell envisaged his medical practice would be both less demanding there, and more prosperous. There was also a real chance of securing a patron. 'Another week passed away, alas! how uselessly,' he confided in his diary. 'Shall I leave this dull place and venture into the vortex of fashion and dissipation at Brighton or shall I not? Prudence, with four children, says stay where you are, but Ambition . . . says go and prosper! What shall I do?'

Since his Lewes practice had been his sole source of income for almost twenty years, the decision preyed on his mind. He told his American correspondent, Professor Silliman of Yale University: 'when I reflect on the many hundreds of families whom, even in my comparatively short life, I have seen reduced from affluence to poverty, I shudder with horror lest such a fate may be mine'. Even his own brother-in-law, Lupton Relfe, had become bankrupt in the early 1830s, and his plight was a continuing concern: 'went with Relfe and saw his poor wife. Heavens! What misery and wretchedness'.

As Mantell's reputation spread, members of the gentry came to visit his museum and this led on one occasion to an introduction to the Earl of Egremont who lived at Petworth House, Sussex. The Earl, a genial man in his eighties, was fascinated by Mantell's collection. It was not uncommon for him to spend several hours, when he visited, browsing through the specimens. Mantell's relationship with the Earl became increasingly cordial during the autumn of 1833, leading to an unexpected development. 'Lord Egremont . . . spoke with me on the subject of my removal to Brighton and munificently offered me a thousand pounds to assist me in the removal!' With such support, how could he fail?

During the autumn, Mantell and his wife took the carriage to Brighton many times to look for somewhere suitable to live. In November 1833, they were shown an imposing house on the Old Steyne, a fashionable part of Brighton close to the sea. Number 20 was a bow-fronted Georgian house, spread over five floors and most prominently placed in the town, barely a hundred yards from the Royal Pavilion.

From every window at the front of the house, Mantell could glimpse the splendour of the palace, the exterior adorned with such a number of oriental domes, towers and turrets that their sheer abundance seemed surpassed only by their redundancy, and spoke volumes of the vast wealth to be squandered by the palace's inhabitants. Beyond, he could see gentlemen's carriages making their way to the palace stables, a great domed building that could house sixty horses. Towards the sea front, smart carriages were lined up by the Old Ship Inn and the Palace Hotel, their rich owners idling time in the cobbled lanes. Directly in front of the house were ornamental gardens stretching across to an elegant row of double-fronted houses, principally owned by the gentry. If he placed his museum of fossil reptiles here, he thought, in the heart of this fashionable part of town, it could hardly escape notice.

'I almost dread to have you . . . withdrawn farther from the Tilgate Forest, the Westminster Abbey of the Old Saurians,' Professor Silliman wrote from America when he heard of the plan, 'for I fear that Science will suffer.' But Mantell was full of optimism as he signed the lease, and was soon 'all in a bustle of removal to Brighton'.

Little by little, the elegant house in Brighton with its marble fireplaces and decorative carving was filled with the eerie relics retrieved from the Age of Reptiles. It was the first museum in the world to show the three known giant land reptiles, and Mantell ordered many expensive new display cabinets. In the most commanding place in the largest room on the first floor he placed the bones of the *Iguanodon*, *Megalosaurus* and *Hylaeosaurus*. Fossils of ancient plants, including ferns and cycads, illustrated the 'Country of the Iguanodon'. There were also numerous cases of chalk fossils, including his fine collection of fishes and casts of the mammoth and mastodon from Paris. Just before Christmas, with his museum complete, his family joined him in Brighton: 'Farewell for ever to Castle Place . . . So ends 1833: and I begin the world de novo!'

The events that followed surpassed even Gideon Mantell's expectations. They had scarcely finished celebrating Christmas – in some disarray, as their belongings were still being unpacked – when arrangements were made for him to give two public talks on geology, which, he confided to Silliman, 'were the means of introducing me to the first people of the town'. The audience was astonished, many learning for the first time of the evidence of former worlds. 'We must read the records of creation in a strange, and perhaps repulsive language,' Mantell said, as he showed them some of the giant *Iguanodon* bones. 'But once this language is acquired, it becomes a mighty instrument of thought . . . in the shapeless pebble that we tread upon; in the undefined mass of rock or clay the uninstructed eye would in vain seek for novelty or beauty; like the adventurer in the Arabian story he finds the cavern closed to his entrance . . . until the talisman is obtained that can dissolve the enchantment and unfold the wonderful secrets that have so long lain hidden.'

The *Brighton Herald* reported that Dr Mantell spoke 'in a style of impassioned and brilliant eloquence to which we have not the ability to do justice . . . All that can be effected is to give the substance of his observations; the fluency, the fervour, the affluence of mind, the command of language, the force, beauty and variety of illustration.' To

Mantell's great delight, when his museum opened, on Tuesdays, 'hundreds of the nobility and gentry flocked through the door'. He felt confident that he couldn't fail to prosper.

Within weeks, he recorded in his diary, 'all the principal persons in this place have called upon me and invited me to their houses, and among many hundreds of acquaintances, I may rank some real friends . . . My museum has already been visited by upwards of 1000 persons.' He told Professor Silliman that his lectures 'were well attended and passed off very agreeably and my society was courted by the fashionables: in fact I was the *Lion* of the season' (his italics).

Accolades continued to feature prominently in the local papers. Mantell was portrayed as the 'Columbus of the subterranean world', a star of geology who was to be 'proudly welcomed to Brighton'. By the time the May Day celebrations were in full swing in the streets, he wrote, 'My reception in this town has certainly been very flattering . . . My noble friend, Lord Egremont, whose liberality has placed me beyond all immediate want of money . . . still countenances me in the most flattering manner.'

His success in Brighton soon brought unexpected opportunities. In May 1834 he received a letter from a quarry owner in Kent, a Mr Bensted, whose labourers had unearthed giant fossils in a pit near Maidstone. Bensted had chiselled along the outlines of the bone until he had revealed a 'portion of the skeleton of an extraordinary animal'. The discovery was announced in the London papers and 'gentlemen travelled a great distance to see it', but no one was able to identify the beast.

Mantell made arrangements to travel to Kent in early June. Although it was past five in the afternoon when he arrived at his lodgings in Maidstone, he set off immediately to find Bensted. He recognised the fossils at once: 'the lower extremities of the Iguanodon: a magnificent group'. Watched eagerly by Bensted, his eye took in all the details: there were several limb bones, a series of fifteen vertebrae, pelvis bones such as the ilium, toe bones, ribs, chevron bones and others. Embedded with all this were the highly characteristic *Iguanodon* teeth. Here, for the first time, were *connected* parts of the *Iguanodon* skeleton.

Any hopes Mantell entertained of taking the remarkable specimen back with him to Brighton were quickly dispelled. Bensted wanted to make a substantial sum out of his find, and several weeks of frustrating negotiations ensued. Eventually, Mantell wrote: 'My *very, very* kind friends, Horace [Horatio] Smith and Mr Ricardo took upon themselves to obtain it, if possible and present it to me.' Within a few weeks, to Mantell's delight, the *Iguanodon* arrived in Brighton. 'Now for three months' hard work at night with my chisel; then a lecture! I must do something to merit such kindness.'

Working late each evening, long after the last carriages had left the Royal Pavilion and the visitors had retired to their hotels, Mantell chiselled away. As the shape of the lower limbs of the *Iguanodon* gradually emerged, he reported to Professor Silliman: 'there are many bones which were not visible when I wrote to you . . . I am now certain that the hind feet of the Iguanodon were very large, flat and enormously strong . . . A femur of the Iguanodon which I have been able – although it was broken into a hundred pieces – to repair and make quite perfect is three feet eight inches long, although shortened somewhat by compression.'

Since the teeth were buried with the other bones, there could be no doubt that they were all part of the same creature. For nearly a decade Mantell's classification of bones had been based on conjecture. At last he had a blueprint of several important parts of the *Iguanodon* skeleton and could confirm his ideas. He also tried to estimate the size of the beast more accurately, measuring the dimensions of corresponding bones in an iguana so as to construct a table showing their proportions.

A comparison of the sizes of the clavicles and teeth of the two creatures suggested the *Iguanodon* might have reached 100 feet long. 'In truth, I believe that its magnitude is here under-rated,' Mantell wrote. 'Like Frankenstein, I was struck with astonishment at the enormous monster which my investigations had called into existence!'

Without a complete skeleton to prove the size of the *Iguanodon*, these comparisons to the iguana lizard were the best guide available. Mantell recognised that if the *Iguanodon* was shaped more like a crocodile, which has a shorter tail than a lizard, 'its total length would, of course, be much

Bones	Recent iguana	Iguanodon	Estimated length of the Iguanodon indicated by the comparison
Teeth		Exceed the recent iguana by 20 times	100 feet
Horn	0.25 inch high	4.5 inches, 18 times larger	90 feet
Os tympani	0.6 inch high	6 inches, 10 times larger	50 feet
Clavicle	1.5 inches long	30 inches, 20 times larger	100 feet
Femur	length of bone 3.5 inches	4 feet, 15 times larger	75 feet
Tibia	2.8 inches	31 inches, 11 times larger	55 feet
Claw bone		16 times larger	80 feet

Mantell's estimates of the size of the *Iguanodon* from his table in *The Geology of South East England* by G. A. Mantell (1833), p. 312.

less than is here inferred'. From the *Iguanodon*'s very substantial finger and toe bones he conjectured that the ancient reptile was much more 'bulky' and thickset than existing lizards. These tentative threads of evidence suggested that the ancient reptiles may have had different proportions from modern lizards, but there was no way of obtaining proof of it. Most of the caudal vertebrae were missing, so he could not prove the length of the tail. Nor did he have bones of the skull or jaw, making it impossible to deduce the shape of the head. Bones of the hands, ankles and crucial parts of the hip bones were also absent. Despite this, Mantell tried to work out the appearance of the creature and even made a first provisional sketch. In his free-hand drawing he depicted *Iguanodon* as a crouching, four-footed lizard, resembling the iguana in shape and proportions.

Such was his enthusiasm that, in September 1834, within a month of receiving the specimen, Mantell was able to enter in his diary: 'finished chiselling out the Maidstone Iguanodon and placed it in the Museum. How it is ever to be got out again, Heaven only knows!' The large rock in which many of the bones were embedded became nicknamed 'the Mantell-piece' by his Brighton friends Horatio Smith and Moses Ricardo, and was manoeuvred next to the other *Iguanodon* bones in the main room.

In scientific circles, Mantell was now so highly regarded that numerous academic distinctions were conferred upon him. Since he had been unable to attend university he was particularly proud to receive a degree from Yale College in America. Professor Buckland dubbed him

The Maidstone *Iguanodon* which Mantell placed in his museum in Brighton and which became known as 'the Mantell-piece'.

the 'Wizard of the Weald' following his spectacular success in inter-preting fossil reptiles. Finally, in 1835, Mantell was presented with the highest award of the Geological Society: the Wollaston Gold Medal. In the history of the Society this medal had been conferred only once before, when William Smith, the father of stratigraphy, was belatedly honoured in 1831. The prize was given to Gideon Mantell 'for the discovery of two genera of fossil reptiles, Iguanodon and Hylaeosaurus'.

That evening, Charles Lyell took advantage of his position as Presi-dent of the Geological Society to promote his friend at the honorary dinner:

> It is now nearly twenty years before I had the good fortune to
> become acquainted with Mr Mantell, and even then my friend
> . . . foresaw some of the results which have since been
> realised . . . His Collection is, of itself, a monument of original
> research and talent . . . an assemblage of treasures which the

Cartoon of 'A Saw-rian' in Mantell's museum by Thomas Hood.

mere industry of a collector could never have brought together
. . . It required his zeal, inspired by genius . . . to bring this to
light, and call into existence those huge Saurians . . . with whose
names we have been made as familiar as with those of our
domestic animals, and which have obtained as real an existence
in our imaginations as if they were living at this moment, in the
Nile . . . Gentlemen, the health of Dr Mantell, the Wollaston
Medalist!

Reports in the local papers show that this accolade was followed by 'loud
and continued cheering'. It was a high point for Gideon Mantell. 'The
past few months have been the most splendid in my existence,' he wrote
in his diary, 'and if fame and reputation could confer happiness, I ought
to be happy . . . I feel I have had very many blessings bestowed upon me
by the Author of all good.'

But all this time he had been neglecting his medical practice.
Paradoxically, his deep interest in science counted against him in
Brighton. Damaging gossip was beginning to circulate. People became
suspicious of signing up with a doctor who, it was said, was more
committed to geology than to medicine and had no time for his patients.
Mantell feared that these rumours were started by rival practitioners, a
view shared by Professor Silliman: 'I am truly grieved that you are so
much disappointed professionally in Brighton, but can easily understand
how envious rivals may make use of your zeal and of your success too, in
science to excite a prejudice.'

As the months went by, for the first time in his life money was fast
becoming a constant preoccupation. Their old home in Lewes was put
up for auction but not one bidder appeared. Eventually, he found a
tenant to pay £60 a year, but this scarcely eased his financial burden since
his Brighton house cost £350 a year and was rapidly draining his few
resources. He had chosen this imposing home in the hope of making a
commercial success of his museum. To cover costs, he had planned to
charge visitors a fee. However, at the last minute, friends advised otherwise: 'science should be cultivated for its own sake', and an entrance

fee would count against him when developing his medical practice.

'So here I am, confessedly one of the most successful practitioners in the county . . . with more reputation as a man of science than I deserve and yet *without a patient!*' he confided to Silliman, six months after moving in. Even when fifteen hundred people had visited his museum, only a few became patients. Far from being a fashionable Brighton doctor with clients in the highest circles, he now found that the prospect of financial ruin was beginning to form an ugly shape in his mind. 'My practice here is very unpromising,' he wrote helplessly in his diary, in June 1835. 'Host of visitors – but *No patients!* What am I to do I know not!'

Gradually, this began to affect every aspect of his life. He became a prisoner of the house, anxiously waiting in for prospective patients. Days passed when he was unable to explore the quarries. He felt obliged to turn down invitations to scientific events, such as the meetings of the new British Association for the Advancement of Science. Whereas in Lewes, people had accepted his scientific interests and his endless excursions had not counted against him, in Brighton he could not be seen to put geology first. He allowed himself short trips to London principally to visit his broker in the City and to sell his stock: £200 in July, and again in the December of 1835.

There seemed no easy way out of the impasse. He could not go back to Lewes, his medical practice was sold and the new occupants had settled in well. He thought of other schemes: he could try to make his name as a lecturer, or raise money to purchase a medical practice elsewhere. If all else failed, he could put his family in lodgings and take a post as a naturalist on a voyage overseas. His hard-earned money was slipping away fast, around a thousand pounds a year since he had arrived in Brighton. He felt himself to be 'floating still on the sea of circumstances'. The financial difficulties added to the tensions within his marriage. 'Very unhappy and unsettled. Alas, I have not found the path of peace,' Gideon Mantell wrote in July 1835. 'Oh that this weary existence would terminate.' Furthermore, it seems likely that Mary Mantell was not prepared to obey her husband meekly, as would have been expected of her, and the uncertainties became hard to bear.

The dreaded possibility, which kept intruding on his thoughts with an ever-increasing urgency, was that he must sell his collection. The giant reptilian bones, the perfect chalk fishes and the exquisite ammonites would fetch a good price at auction. If he sacrificed these beautiful treasures, then he might be able to buy a practice, and that would please his wife. 'I cannot think of your selling your cabinet,' Silliman wrote to Mantell from America, 'it would be almost next to – but it is true at a great distance – *selling your children!*' (his italics).

By December of the same year, something had to be done. Gideon Mantell had attracted a considerable following in Brighton. Horace Smith and Moses Ricardo who purchased the Maidstone *Iguanodon* for him, George Richardson the son of a local draper, and others, rallied round with a scheme to create a 'Scientific Institution' based on his museum. If the public were to donate a shilling for admission, Mantell could be paid a fee, and since there were thousands of visitors to Brighton every year, the institution would surely break even. All this time there had been silence from the palace across the road, but with great enthusiasm, they approached the royal family directly for support.

For all their high hopes, the final plan did not emerge in quite the form Mantell would have wished. They couldn't raise enough money to create a separate museum. It soon became clear that there was no alternative: the new scientific institution would have to be based at his home in the Old Steyne. At a meeting in the town hall to discuss the funding, the situation seemed to spiral rapidly out of Mantell's control. The *Brighton Gazette* reported on an animated discussion in which 'Mr Mantell agrees to give up his house on the Old Steyne – a most eligible situation – at a considerable reduction in rent, for the purposes of the Institution . . . it was unanimously determined that the plan submitted to the meeting should be carried into immediate effect.' Whether Mary Mantell was in the town hall to witness this public disposal of their home, or learned of the 'unanimous' decision afterwards, is unclear.

Bit by bit, a price was put on all the various fragments of their lives. Mantell's home in the Steyne would be let to the new institution for £150 a year; the collection, which occupied most of the principal rooms, was let for a further £250 a year. A room at the top of the house could be spared for Mantell himself. A further room would be needed for a curator. There was simply no space for his family. His wife and children would have to be placed in lodgings elsewhere. Mary found herself forced to acquiesce in an arrangement that turned the family out of their home; the relics of ancient creatures had now taken over their domestic lives entirely.

Christmas 1835 marked a turning-point in the Mantells' marriage. It was impossible to negotiate even the simplest tasks of the day without running aground on the vast submerged icebergs of domestic hostility that their relationship had become. 'One of the most miserable Xmas days I have ever spent,' Mantell wrote. 'What misery have I not endured this year! . . . My prospects are so cheerless; with "none to bless me, none whom I could bless", oh how my soul yearns for some kindred spirit on whom it could lavish all its tenderness!' While the children prepared for the delights of Christmas, Mary Mantell was unforgiving in her anger and quite unconsolable. For her, the plight they were in was all his fault; it was his reckless pride and his selfish pursuit of his own interests that had brought them to this hopeless point. Her condemnation, on top of all the other difficulties, utterly defeated Mantell: 'Gracious being, Oh enable me to bear up under the miseries that surround me on every side. Oh take me from a world for which I am so wholly unsuited.'

After Christmas, gentlemen arrived at their house to draw up formal proposals to turn the home into an institution. They soon received word from the palace: 'their Majesties cannot allow their names to be placed as sanctioning the undertaking'. Undeterred, Horace Smith and Moses Ricardo went to see Lord Egremont to request his name as patron. When they returned on 4 January, eager to tell Mantell that His Lordship had generously given yet another thousand pounds, the news that would have been so welcome a few years ago now failed to raise his spirits. 'Worn out with care and fatigue,' he wrote despairingly.

As Mary Mantell struggled to rearrange her life around the new plans, the enthusiastic reports in the local papers were, for her, hurtful public pronouncements on the failure of their private lives. 'It is with feelings of unmingled satisfaction that we allude to the brilliant prospects of this yet infant Institution . . . We have much pleasure in learning that the arrangements . . . are now nearly complete, the new cases are almost finished,' enthused the *Brighton Herald* in March 1836. According to the *Gazette*, 'Dr Mantell's museum . . . is more interesting and perfect than any other in Europe.' One local estimate suggested there were nearly thirty thousand specimens. A portrait of the munificent Earl was hastily commissioned for the town hall.

The next month, Mary Mantell took her drapes, china and furniture to lodgings in Southover Street, Lewes, and moved in with the children. From her modest cottage at the bottom of the hill she could see their former home at Castle Place, a daily reminder of the hopeful early years of their marriage. Some historians have suggested that she encouraged the children in antagonism towards their father. The older ones made plans to leave home. Walter was apprenticed to a surgeon, and wanted to emigrate to New Zealand once he qualified. Ellen talked fancifully of moving to America. The two youngest, Reginald and Hannah, were still away at school.

Once the family had moved out, the doors of the Sussex Scientific Institution and Mantellian Museum were formally opened to the public. George Richardson, 'a man of powerful intellect' according to the *Brighton Guardian*, blossomed as the newly appointed Curator. Mantell was given a room in the top of the house where he could await patients and attend to the business of the museum. His whole life seemed to him to have revolved around the petrified remains of former worlds, but now it had begun to pall. 'In truth,' he wrote, 'I am now sick of the cold-blooded creatures I am surrounded by.'

The next month, when Mantell went to visit fourteen-year-old Hannah in Dulwich, on the outskirts of London, he was concerned to find the school had not informed him that she was severely ill. 'She was attacked with disease of the hip joint while at school and unfortunately the earliest symptoms of that infection were mistaken for a mere common rheumatism,' Mantell told Professor Silliman. Not satisfied that she was receiving the proper treatment, he anxiously returned the next day to see what else could be done. Over the next few months he visited frequently, noting any improvements in his diary and making plans, as her strength returned, for outings to galleries or on picnics.

Both Gideon Mantell and George Richardson felt strongly that the fate of their new Scientific Institute in Brighton depended 'on the success or failure of our attempts to cultivate a taste for scientific knowledge in the town'. They devoted themselves tirelessly to finding ever more original ways to engage an audience. Mantell's talks became increasingly popular; on one occasion eight hundred people crammed into the town hall to hear him speak. It was not uncommon for eminent visitors to attend – such as Michael Faraday, who was making his name with his studies of electromagnetism at the Royal Institution; Louis Agassiz, a Swiss naturalist, who was astonished at the 'beauty and perfection' of Mantell's chalk fossils; Professor Buckland, Roderick Murchison, Charles Lyell and others from the Geological Society. 'As a lecturer, Mantell had no rival and could hold his listeners spellbound,' announced the *Herald*. 'He was even more masterly and eloquent than before', according to the *Gazette*. 'Loud and continued applause evinced the high intellectual gratification which Dr Mantell had afforded to his distinguished auditors.'

Despite their estrangement, his wife still tried to support him; to his great delight, he would occasionally see her in the audience, with one or two of the children. In fact, his domestic situation must have seemed more hopeful, because in July 1836 he wrote optimistically, 'Happiness may yet be the lot of those I love. My sweet Hannah Matilda decidedly better', and 'Mary very happy and kind. Could I but find a good professional opening, the more labor the better, and all would be well.' Shortly after this, Mary moved back to Brighton to be nearer to him, in a little cottage in Western Road.

Mantell could reassure his wife, at last, that he had some patients in Brighton, and that as the Institution prospered he would raise the money to buy a new practice. By Christmas, Hannah was well enough to attend a concert with the family. 'I am grateful to the Eternal for the blessings he has permitted me,' he wrote in his diary in the December, delighted at Hannah's progress. 'A gipsy foretold me that 1837 would make or mar my fortunes! Be it so, I am prepared for good or evil.'

In the spring of 1837, at last the tide of fortune seemed to have turned in their favour. Mantell's talks were beginning to attract interest well beyond Brighton. According to the *Lancet*, his 'Popular Lectures on Physiology' were illustrated by drawings 'which although anatomically correct, were deprived of every repulsive character . . . Ladies of rank and fashion were seen handing round glasses containing dissections of the eyes of sheep, oxen & etc and examining them with as much interest as the contents of caskets of jewels often excite.'

To maintain interest in the museum, Mantell and Richardson devised ever more ingenious publicity. In June 1836 the *Herald* had reported 'An Extraordinary Occurrence', in which Richardson was disturbed by a noise in the museum cabinets: 'on turning, Mr R saw with horror the whole collection of these gigantic bones in motion! The thigh bones had placed themselves erect and were dancing about as though looking for legs; the head glided towards the trunk . . . the jaw-bones were clanking together inviting the teeth from the other end of the room and claws came from under the table.' The ancient beasts, reassembled, devoured everything in sight, 'making a desert of the umbrellas, a brand new hat, an edition of Dr Johnson and a complete file of *The Times* . . . This wonderful reanimation will be discussed at the next *conversazione*.'

Not surprisingly, all this took its toll. By day Mantell was involved in the museum, and the evenings were passed in discourses and *conversaziones*. While the newspapers continued to praise his 'highly lucid and delightful style', Mantell felt 'wearied to death'. The effort of public performances while his private life was still in such turmoil was becoming too much for him: 'very wretched . . . I tried to begin my lecture, but could not succumb to my wishes.' In spite of his financial troubles, not all the money raised was for the benefit of the Institution: 'Gave a lecture at the Old Ship – 350 persons present – clear profits 25 pounds, for poor Phillips, the florist, who is quite blind.'

By spring 1837 they had at last succeeded in getting royal patronage, and the name of the museum was formally changed to the Sussex Royal Institution. But despite this honour, there was still no money forthcoming from the palace – which they could see from the museum windows, the chandeliers twinkling until late, illuminating the very visible signs of wealth of which none, apparently, could be spared to help them. As the months passed, despite their superhuman efforts, the Sussex Royal Institution was losing money fast.

'How singular is my present position,' Mantell wrote to Silliman. 'Popular as a lecturer, my society courted by the first in rank and in science, most successful as a practitioner . . . for twenty years; with a museum of my own equal to, if not surpassing, any private one in Europe . . . patronised by one of the most wealthy noblemen in England . . . yet I am in the present moment in the greatest anxiety as to my future prospects. Envied by many – Alas! How little can the world judge of our real state.'

Hannah seemed sufficiently well in February 1837 to return to school. But two months later, Mantell brought her home again, greatly concerned at her condition. All the while, financial problems were becoming more pressing. He seemed unable to raise the money to reunite the family and was becoming increasingly uncertain whether his wife even wanted to come back to join him. 'The day drags on, though storms keep out the sun; and thus the heart will break, yet brokenly live on!' he wrote in his diary, 'I am resigned to the will of Him who knows what best His creatures can bear.'

Gideon Mantell knew that there was now no escaping the prospect he dreaded most: selling his museum of giant fossil reptiles. All his life he had been building up his collection; he had begged for it, saved for it, travelled hundreds of miles for it, sacrificed domestic comfort for it. He couldn't quite believe it would come to this, and for a while clung to the belief that there might be some last-minute reprieve. Sometimes, stubbornly ignoring their financial plight, he still couldn't resist making new acquisitions at auctions: a large portion of the skeleton of *Hylaeosaurus*, some new *Iguanodon* or mastodon fossils. He wrote in disbelief to Professor Silliman, 'I am obliged to sell my Museum!!! To be compelled to take this step is you will readily believe a severe trial to me, but I find it must be so, either medicine for a living, or science, one must be renounced. And so like Shakespeare's apothecary, my poverty not my will consents.'

As the sale became inevitable, once more local friends rallied round with a new scheme. They aimed to raise £3,000 by issuing a series of shares in the museum. The obliging Lord Egremont promised that he would purchase £500 worth of shares, once they had sold a sufficient number. Many supporters stepped forward: 'Mr Ricardo, himself a subscriber of 100 pounds has handed in the name of his brother for 50 pounds,' said the *Gazette*. 'Miss Wright, the first female subscriber has taken two shares of 50 pounds . . . it is determined to apply immediately to the Noblemen and Gentlemen of the County in order that the requisite sum can be raised.'

All the while, Gideon Mantell persisted in his valiant attempts to attract more supporters. Over six weeks, in a series of lectures, he delineated an entire history of the earth. Starting with the human epoch, each week he ventured further back in time, revealing the geological evidence for a succession of different eras. His second talk featured Cuvier's discoveries: 'the period immediately preceding the appearance of Man on the earth, and the large Mammalia such as the Mammoth and Megatherium who constituted its chief inhabitants'. He also described Buckland's cave work, explaining that the hyena and bear were contemporaries of the mammoth. Probing ever deeper into the earth's crust, in his third lecture he discussed Lyell's studies of the Tertiary layers, describing the different types of shells found in the London and Paris basins. The Age of Reptiles entombed in the Secondary strata below the Tertiary formed the subject of his fourth and fifth talks.

It was his life's work, and as he displayed some of the remarkable fossils he had found, each giant reptile was vividly conjured to life. 'The prejudice that once existed against the supposition that creatures of this nature were once the chief inhabitants of the earth is now passing with the advance in knowledge, and the geologist is no longer subject to reproach for stating a fact which is proved by cumulating evidence,'

Mantell declared. Like his friend Lyell, he did not support the evolutionists. He reassured his audience that the extraordinary beings that once thrived on the earth 'were forms of happiness which were precisely adapted to the condition in which they were placed'.

Several hundred people flocked to the town hall to hear his last talk, on Saturday 21 October 1837. The clock in the tower struck three; the parked carriages could be seen extending all the way back up North Street. Inside the hall, people were crammed in at the very back and in all the doorways. The tickets, at two shillings and sixpence each, were sold out. This time, Mantell delved even further into the past, describing the discoveries made by Murchison in the Silurian rocks of the Transition era: the strange trilobites, crinoids, echinoids and corals 'found in very early formations of the earth'. In his inimitable way, he created a dramatic picture of invertebrate life 'in the earliest seas'.

He was coming to the end:

> I cannot perhaps more appropriately close my present discourse than in the beautiful stanza in which Lord Byron apostrophises the sea . . . in language which is as scientifically just as it is poetically eloquent and beautiful:
>
> > Thy shores are empires, changed in all save thee . . .
> > Time writes no wrinkles on thy azure brow,
> > Such as Creation's dawn beheld, thou rollest now.

As he finished, the entire house stood, and the hall echoed with resounding applause.

But any hopes for a reprieve were short-lived. Three weeks later, Gideon Mantell heard with a sense of shock that Lord Egremont was dead. In December 1837, the planned anniversary dinner of the Institution gave way to a hastily arranged emergency meeting. Mantell's tribute was 'in a style of eloquence surpassing even his usual style of oratory', according to the *Gazette*. Yet, even as he spoke, joining in the chorus of resilient, hopeful voices forming plans for the next year, it was obvious to him that this was the death knoll. There was no hope now of saving the museum.

Mantell's collection was offered to the Brighton Council for £3,000, although it was estimated to have cost him over £7,000 to gather together all the specimens over the years. But Council officials turned down the scheme. And the new Lord Egremont had no interest in the museum. Mantell realised he would have to disperse the collection and try to sell pieces where he could, the outcome he most dreaded. 'How I wish you could see it before it is dispersed,' he told Silliman, in September 1837. He begged him to see if he could find a purchaser in America. But no buyer could be found, even at a reduced price, for this, the finest collection of giant land reptiles that had yet existed.

On 30 December, the *Herald* carried a short announcement: 'We learn with the deepest regret that the dispersion of Dr Mantell's museum is now inevitable.' It urged the reader not to let this happen: 'a very strong feeling we know exists in the town against the discontinuance of the establishment . . . That our county, abounding in noblemen and gentlemen of wealth and intelligence should allow a treasured collection such as this to be dispersed would constitute a disgrace!'

But the entreaties were to no avail. Early in 1838, Mantell became resigned to the only other possibility. He sat at a desk in the corner of the museum, surrounded by those so familiar shapes that now inhabited the house and imbued it with an unearthly stillness. He could hear only the odd noise outside, as though coming from a great distance – a horse and carriage passing, the occasional seagull. Slowly, he began to write to his old acquaintance Charles Konig, at the British Museum.

My dear Sir,
I am desirous of entering into a treaty with the Trustees . . . Tell me who they are, and through whom I shall best succeed. [To pressure them in his favour, he added somewhat untruthfully] I am besieged with applications from local institutions; but I am now resolved to have the Collection in the British Museum . . .

Anxious to secure a sale, he felt obliged to explain his straitened circumstances to other senior figures in geology, who also draughted letters to
the British Museum on his behalf. The new President of the Geological
Society, Roderick Murchison, wrote: 'such is the real value of the collection that it would do honour to the nation if placed at . . . the British
Museum . . . I should consider it a grave imputation on our national
character if this opportunity be lost!'

But the museum officials were not moved – they had a growing
number of amateur collections to consider. A month later, on 17
February, Mantell wrote again:

> The Collection consists of many thousands of specimens, but the
> grand features, and on which I rest its claims to your attention
> are the remains of the Iguanodon, Hylaeosaurus and other
> colossal reptiles and fossils peculiar to the Wealds of
> S. E. England . . . The sum for which I offer my Museum to the
> National collection is 5000 pounds . . . I am willing to enter into
> any arrangement or negotiation which you, my Lords and
> Gentlemen, may propose.

The appeal was of no use. Correct procedures had to be followed, he
was told, to establish the merits of the collection; a parliamentary grant
would be needed for an application of this size; recommendations from
other geologists were also required, to assess the value. Soon, officials at
the British Museum received more letters begging assistance: Charles
Lyell, Lord Northampton, Roderick Murchison and Adam Sedgwick all
petitioned on Mantell's behalf. But they were no more successful than
Mantell had been: there was no money for a purchase of this sort this
year. Eventually, Lord Northampton and Roderick Murchison went to
Westminster to appeal directly to the Chancellor. New and seemingly
insurmountable obstacles were uncovered when it was realised that the
annual funding estimates for the British Museum had already been
submitted to the Treasury. Additional funds could not possibly be
committed until next year.

In the end, after weeks of negotiation, an agreement was reached in London. The British Museum would take the collection, but Mantell would have to wait a year for the money. There was also the trifling matter of the price to be settled. Professor William Buckland stoically rallied on Mantell's behalf, advising the trustees to pay the sum requested by Mantell as he had 'great confidence in Mantell's accuracy of judgment in such matters'. Testimony from one as eminent as Buckland might well have clinched the matter.

However, it did not. Negotiations were becoming so protracted and complex that Mantell was beginning to wonder whether he would be obliged to sell his collection 'by the hammer'. Then, during the autumn of 1838, senior officials of the British Museum including Charles Konig and Henry Stutchbury came down from London to value the items for themselves. All agreed on the difficulty of pricing so unique a collection.

Mantell and George Richardson waited patiently while officials checked and cross-checked catalogues and discussed the merits of the specimens. Because of the sheer number of fossils, even the highly skilled Stutchbury could not number the material in the way considered most fitting in the time available. Mantell offered to help with the giant reptiles and the chalk fish fossils, his two favourite sections, but became too 'distracted' to get involved. Eventually, with the assistance of Richardson and a 'careful lad', the lists were completed, and the grand sum of £4,087 was agreed. On Mantell's behalf, the Marquis of Northampton arranged a post for George Richardson at the British Museum, as a sub-curator. In November 1838 the *Gazette* announced, 'The Mantellian museum closed this week and will be transferred as quickly as possible to the British museum . . . spring vans have been engaged for the carriage of the specimens . . . It is expected that the whole will be removed by the 7th of next month.'

In December Gideon Mantell returned from London, where he had been making arrangements to secure a new medical practice, to take a last look at his collection: 'What a lesson of humility! What a proof of the vanity of human expectation,' he wrote in his diary. All the fossils were packed in boxes and stacked, waiting. His whole life seemed to be in transit. Mary had not gone with him to London, where he had plans to start a practice in Clapham. As he paced the empty floors, he wondered if he could persuade her to return to him. If she came now, she would see that he was giving up his beloved geology for good. Everything was packed; everything was going. Not one fossil, not one treasure had he kept back for himself. If only she could see how he had quite deliberately deprived himself of even a memento, she would understand what he had given up.

But he was quite alone in the house in Brighton. There was no sound, apart from Richardson busying himself upstairs among the cases. He caught sight of the family motto engraved on the coat of arms above the door: *Nil desperandum*, 'Despair of Nothing'. His predecessors, too, had lost everything they valued.

When he woke the next day, he could hear the sound of horses outside. He quickly drew back the curtain of his small attic window, and saw that it was not his wife arriving. There was mayhem below in the Old Steyne. Some ninety horse-drawn vans had been ordered to take the entire collection. Museum staff were beginning to stack up boxes in the street. Coachmen were awaiting instruction. It was commotion.

Eventually they packed the last van bearing the *Iguanodon*. Mantell watched as the entourage set off down the Steyne, turning left at the end of the Green for the London road. He heard the sound of the horses' hooves receding into the distance. He had given up his *Iguanodon*, the symbol of all his youthful hopes of success.

While he was trying to establish the planned medical practice in Clapham in the spring of 1839, his wife finally deserted him. During that summer, his older children also made arrangements to leave. Ellen, at twenty-one, was old enough to leave home in her own right. Walter, having finished his training as a surgeon in Chichester, was determined to emigrate to New Zealand. Despite strong opposition from Mantell, who wanted his son to settle in a medical practice nearby, Walter finally set sail on 15 September. Mantell was struck with the awful realisation that he had lost everything that he valued in life: his hopes for a scientific career, his precious collection, and his family. Overcome with disappointment, he deleted his diary entry for the day: 'My son Walter, and my daughter Ellen . . . [words blotted out]'

Gideon Mantell was not quite alone. His daughter Hannah was at home, and in the summer of 1839 she seemed so improved that they began to make plans for after her recovery. He knew she would always suffer some lameness, but they could surely stop the spread of the infection. 'My sweet girl is still wholly confined to her bed, but she is better than she was, and does not suffer. She is obliged to lie constantly on her back, but she is still able to draw, paint, knit, write and work, and her sweet temper and disposition make everything delightful around her.' He was pleased with her progress. 'How mysterious are the ways of Providence!' he told Silliman. 'If there was ever a human being free from the waywardness of temper and the usual failings of mortals it is that sweet girl! Is it to teach us that by patient suffering we can alone be made perfect?'

In the autumn, with the colder weather the infection in Hannah's hip flared up once more and she became weak again. Mantell would bathe and nurse her wound for an hour each morning and evening, trying to conceal the anxiety he felt for her. Her hip was now so wasted by the infection that the bone was painfully prominent through her skin. He was mortified by his inadequacy as a doctor, by the fact that he could not help the person who was dearest to him. He made her an invalid carriage and became so anxious for her that he never left her, except when called out professionally. He moved to the bedroom next door so that if she called out, he would hear her, day or night.

One evening the servant summoned him. Hannah had fainted from a sudden and very severe haemorrhage. Mantell attended to her as best he could; his sister arrived, and his niece, to help with the emergency, and for a few days Hannah seemed to rally. But early one morning she collapsed again. She asked her cousin to summon Mantell. This time, the haemorrhage was so severe that before he could step inside the room she was insensible. A few minutes later 'her gentle spirit passed away'.

> My sweet girl, Hannah Matilda suddenly expired from haemorrhage after a long and distressing illness of three years' duration:

and thus [words blotted out] one whose sweetness of disposition and affectionate heart endeared her to me beyond even the natural ties that united us, is taken from me!

> Before the Chastener humbly let me bow
> O'er hearts divided and o'er hopes destroyed!

All that remained was a little box of her treasured possessions: a story she had written, a coronation medal, several trifling ornaments and a piece of embroidery that she had just completed for Mrs Silliman. A few days later he interred the remains of his most cherished daughter in the cemetery at Norwood. 'In a state of depression almost unbearable,' he wrote to Professor Silliman. Many months were to elapse before he was able to write in his journal, 'have in some measure recovered my tranquillity of mind'.

PART THREE

I I

Dinosauria

All things bright and beautiful,
All creatures great and small,
All things wise and wonderful,
The Lord God made them all.

Mrs Alexander, 1848

Eighteen thirty-seven, the year the young Princess Victoria was proclaimed sovereign of the British Empire, was also a landmark year for Richard Owen. At the age of thirty-three, he won a coveted prize that would earn him a special place for generations in the history of science.

Among the leaders of the British Association for the Advancement of Science there was a growing concern that while the French had produced a Cuvier, and in Switzerland the naturalist Louis Agassiz was doing masterly research on fossil fishes, British science was losing out as foreigners gained access to the spoils of British discoveries. The giant fossil reptiles were seen as uniquely British. Yet there was no prominent British scientist held in sufficiently high esteem to be charged with the task of interpreting and classifying them. The leaders of science in London had no doubt that this was a field in which Britain should excel. They needed a hero who could be swept up and placed on a pedestal, someone like Baron Cuvier.

It was Richard Owen, not Gideon Mantell, whom the establishment chose to be their figurehead. Even though Mantell had discovered the *Iguanodon* and the *Hylaeosaurus* and distinguished himself with studies of the giant reptiles, it was Owen who won the backing of the BAAS to draw up a 'Report on the present state of knowledge of the Fossil Reptiles of Great Britain.' His triumph was a measure of his political as well as his scientific skills.

Although one of the aims of the BAAS was to promote science in the provinces, which in theory should have benefited Mantell, in practice, the London scientific gentry invariably hijacked the proceedings. In the early years of the BAAS, members of the aristocracy far outnumbered professional members as past presidents. Owen was highly visible to such worthy gentlemen through his contacts at the Royal College, while Mantell, struggling to earn a living in Brighton, had failed even to attend the meetings. Owen's scientific reputation, too, was enhanced by his studies on fossil mammals brought back on the *Beagle* by Charles Darwin. His treatise on the comparative anatomy of teeth was eagerly awaited in scientific circles.

Owen had one other trump card. His father-in-law, William Clift, was on the three-man grant committee of the BAAS, which voted him £200 in 1838 with which to begin his research. George Greenough, who had clashed with Gideon Mantell years before over the interpretation of Weald strata, was also on the committee. Curiously, Charles Lyell was the third member. It is possible that Lyell felt his friend Mantell's personal circumstances were too chaotic for him to carry this work forward, or perhaps he found his hands were tied, since Owen had been specifically recommended for the task the previous year.

Richard Owen began his survey of fossil reptiles by reviewing the 'Enaliosauria', the sea lizards such as ichthyosaurs and plesiosaurs, many of which had been found by Mary Anning. These marine lizards held a special interest for Owen since Geoffroy Saint-Hilaire, in relentless pursuit of evidence for 'progressionism', had proposed that crocodiles might have developed gradually from ichthyosaurs. Owen seized his opportunity, confident that he would soon make the Frenchman's assertions look ridiculous.

He travelled to meet private collectors such as the eccentric Thomas Hawkins at Sharpham Park, Somerset, who had been fascinated by the marine lizards for years. Hawkins had spent his inheritance extravagantly trying to obtain the very best specimens from Lyme, once even paying to 'throw down as much of the cliff as was necessary' to obtain an ichthyosaur. He had acquired some of the largest fossils, including a superb *Ichthyosaurus platydon* over twenty-five feet long, many of which were sold on to the British Museum during the 1830s. 'Hawkins has done some wonderful work disencumbering the old Saurians from their stony shrouds,' Owen told his father-in-law. After his visit to Hawkins, Owen decided 'to take a run down to make love to Mary Anning at Lyme and then post home.'

However, his plans to flatter Mary Anning, and doubtless exploit her ideas, do not appear to have come to fruition. At Lyme he met Buckland and Conybeare, who, he told Clift, 'made me a prisoner and drove me off to Axminster, where Conybeare is the rector'. When he did meet Mary Anning the next day, 'we had a geological excursion . . . and had like to have been swamped by the tide. We were cut off from rounding a point, and had to scramble over the cliffs.'

Although Mary Anning's impressions of Richard Owen are not recorded, she was probably more guarded than he would have liked. By now she was fully aware that her discoveries were being exploited by the gentlemen of science, and this sometimes caused resentment: 'She says the world has used her ill and she does not care for it,' wrote her young friend Anna Pinney. 'According to her account, these men of learning have sucked her brains, and made a great deal by publishing works, of which she furnished the contents, while she derived none of the advantages.' Mary Anning told another friend: 'the world has used me so unkindly, I fear it has made me suspicious of everyone'.

Despite her endless laborious searches, Mary Anning was still struggling to make a living. To add to her difficulties, in the late 1830s she entrusted her life savings of a few hundred pounds, accumulated from the sale of fossils, to a private investor who then disappeared. All efforts to retrieve her savings from the conman failed, and he was never seen again.

William Buckland, concerned at the continuing hardships she faced,

A watercolour sketch of Mary Anning.

tried to raise money on her behalf, and members of the BAAS donated £200 towards her fund. By 1838, the year of Queen Victoria's coronation in Westminster Abbey which cost the nation £200,000, Mary Anning, for the first time in her life, had a certain income of £25 a year. It was enough for a diet of potatoes and bread and would save her from starvation if no new fossils were found.

Owen's scramble over the cliffs at Lyme was one of the very few occasions when he actually set foot in a quarry or on the shore. He had little time for the hazards of collecting specimens, and as the rising star of the BAAS his brief was to exploit the discoveries assiduously made by others such as Anning and Mantell. Indeed, Sir Philip Egerton, head of the BAAS and a Tory MP, wrote to Owen saying that he had so great a regard for his talents that he felt he was the most 'supremely fitted' to gather the 'harvest . . . offered in our Collections'.

Richard Owen did not disappoint Sir Philip and the other gentry in the audience when he read his report on sea lizards at the BAAS in Birmingham in 1839. At first hearing, his study, although technical, was not strikingly original. Much of the work on the anatomy of sea lizards had been carried out by early investigators such as the Reverend William Conybeare. Since almost entire skeletons had been found, there was little to doubt in their interpretations. Owen identified ten species of ichthyosaurs and sixteen species of plesiosaurs. He outlined their characteristics and showed how the backbones and limbs of these lizards had been superbly adapted for marine life.

But to the delight of the squires of science who had gathered to hear their protégé, Owen used his platform to criticise the radical French. 'Do the speculations . . . of Lamarck and Geoffroy Saint-Hilaire derive any support, or meet with additional disproof, from the facts?' he asked. 'We have the opportunity of tracing Ichthyosauri, generation after generation through the whole of the immense series of strata.' But, he mocked, at no point did they gradually metamorphose into crocodiles, as Geoffroy Saint-Hilaire had proposed. 'The very species which made its first abrupt appearance in the lowest strata, maintains its characters unchanged and recognizable in the highest of the Secondary strata,' Owen declared. 'In the chalk the genus Ichthyosaurus quits the stage of existence as suddenly as it entered it . . . and with every appreciable character unchanged. There is no evidence whatever that one species has succeeded or been the result of the transmutation of a former species.'

The leaders of the BAAS were delighted; what they were hearing was nothing less than the work of 'the greatest comparative anatomist living'. Sir Philip Egerton, who had been instrumental in promoting Owen, described the report as 'glorious', adding that he felt 'no regrets . . . of my humble efforts . . . of accelerating the production of so valuable a report'. Owen was promptly offered a further £200 by the BAAS to extend his study of fossil reptiles to include ancient crocodiles, turtles and the gargantuan land lizards named by Mantell and Buckland.

Family archives show that Richard Owen 'spared no trouble' in preparing the second part of his report on British Fossil Reptiles. He was determined to gather all the information he could on their anatomy, to ascertain their shape and proportions more accurately, and to classify them. For this, he could build on an established classification of saurians initially proposed by a distinguished German naturalist, Hermann von Meyer, in 1832. Von Meyer had grouped the carnivorous *Megalosaurus* and herbivorous *Iguanodon* together as 'Saurians with Limbs similar to those of the heavy land Mammalia'. Later, Mantell had added the *Hylaeosaurus* to this group.

Unlike the early investigators such as Mantell, Owen benefited from a number of developments. Not only were there many more fossils to examine in amateur collections all over England, but also the new railways provided easy transport. There was even time to describe the novelty to his wife: 'From Derby to York there are divers tunnels . . . The combination of sounds, rattling along at full speed, the rushing of the rapidly displaced air, and the incessant yell-shriek of the steam-screamer, kept up to warn the tunnellers, defies all description. Pitch darkness, the sparks from the engine darting through the palpable obscure, and the cowering figures like shadows as we swept past them.'

'Since I have left you,' he told Caroline, 'I have gone over more ground than ever I did in my life before in the same time.' In the north, he told her, the museums were 'crowded with visitors – working classes'. It was 'all very orderly and "paws off" . . . but hitherto I have been disappointed by the Saurians'. Soon he came back south, in search of better specimens.

In the City of London he met a wine dealer called William Saull, who had opened a museum displaying many fossils from Wealden strata in the Isle of Wight. On another trip by mail-cart across Sussex he learned of a notable collection at Horsham owned by George Bax Holmes. Holmes had inherited wealth in 1836, given up his job as a chemist, or 'druggist', and devoted his time to geology. It was soon apparent that he had been gathering fossils from Mantell's favourite site in the Tilgate Forest.

For Owen, this was the perfect opportunity to obtain more material from his rival Mantell's territory. He cultivated Holmes, flattering him with his interest and promising that his fossils would be prominently credited in London. Greatly encouraged, Holmes soon put his entire collection at Owen's disposal: 'I hope thou wilst not fail any hesitation in borrowing them as they are most entirely at thy service. I hope that when thou comest in the Spring, thou wilst allow thyself time to examine them more thoroughly.' Owen 'borrowed' a boxful of specimens from him, and received fossils regularly 'by the Horsham coach'.

Of all the collections, probably the greatest asset for Owen was Gideon Mantell's superb fossils of the giant land reptiles at the British Museum. Ironically, Owen had easier access to these than Mantell himself, since Owen lived only a mile away from the museum at Lincoln's Inn Fields. Mantell's collection not only contained some unique fossils to study, but also, with the sheer number of bones and casts he had acquired, provided a wonderful source of reference against which any new finds could be compared. Now a short stroll from his home, the trophies of twenty-five years of desperately hard work had fallen unwittingly into the hands of the man who was poised to turn Mantell's downfall to his own great advantage. Gradually, Owen began to 'reap the rich harvest' of which the leaders of science such as Sir Philip Egerton had sown the seeds.

While Richard Owen was confidently acquiring a name for himself, Gideon Mantell, by contrast, seemed sunk in misfortune. Since his move to London, he was living quietly with his only remaining child, Reginald, who was often away at school. His new medical practice at Clapham Common, purchased from Sir William Pearson, absorbed much of his time. Although his rented house in Crescent Lodge was 'very agreeably situated' on the main Brighton road, he did not venture out much into London Society and repeatedly declined invitations to lecture. As for geological research, that pleasure, he felt, was past. 'I am most anxious about Dr Mantell,' wrote his friend Robert Bakewell in the autumn of

The Country of the Iguanodon as envisaged by Gideon Mantell
and painted by John Martin in 1838.

1839. 'I have not seen or heard of him since his son sailed for New Zealand.'

Not for the first time in the last few years, Mantell was finding that life had lost its charm. 'I have no home for my affections,' he wrote in his diary. He confided to Professor Silliman, 'I have no companion, no one whose smile or approbation would cheer me on . . . There was a time when my poor wife felt deep interest in my pursuits . . . but of late years she was annoyed rather than gratified by my devotion to science.' On the anniversary of his wedding day on 4 May he was quite alone, and 'suffering severely'.

The loss of his daughter Hannah, although he made valiant efforts to overcome his grief, was still more than he could bear. He took the carriage frequently 'to the grave of my departed Angel', sometimes even visiting the cemetery at Norwood twice in one week. The accumulation of disappointments had affected his stamina. He felt 'sadly broken up in health and energy . . . the vigour of manhood is gone for ever'.

236

Ironically, when the content of his last lecture series at Brighton was published in book form in 1838 as *The Wonders of Geology*, it became very popular. The frontispiece showed a dramatic engraving of the 'Country of the Iguanodon' painted by the artist John Martin, showing giant reptiles locked in combat and pterodactyls flying above. Mantell was delighted with the painting, and felt at last that the ancient landscape had been rescued from the 'oblivion of all ages'. Unlike his earlier books, *Wonders* sold well, the first thousand copies selling out within a month. 'My farewell to Geology has therefore been a flattering finale to my labours,' he wrote, 'and I must now be content to sink into the jog-trot of a medical practitioner.'

The evidence suggests that Mantell tried to help Owen gather materials for his 'Report on the Fossil Reptiles of Great Britain'. At one point, in November 1840, relations between them were so cordial that Owen even invited him to dinner with William Buckland. During the evening, Owen showed his two guests his new microscope. 'Richard entertained them to their heart's content,' Caroline wrote. 'They made some experiments with blood globules. Dr Buckland's blood was irregular . . . Dr Mantell . . . on examination proved to have blood globules of a decidedly larger size than the others. Dr Buckland was just saying with that droll look of his, "Why Mantell, you see you have a good deal of the reptile about you," when the news was brought in that the Queen was safely delivered of a little princess, so the discussion was stopped by all the gentlemen drinking to the health of Her Majesty.'

The microscope was not just a diversion – it was also an important new research tool. Later, in his study at the College of Surgeons, Owen prepared minute slivers of the ancient *Iguanodon* teeth and compared them with sections from the teeth of the modern iguana. As he brought the sliver of tooth, millions of years old, sharply into focus down the lens, he could see that it had a different internal structure from the iguana. It struck him that the name '*Iguanodon*', or 'iguana-tooth', was inappropriate. Sections of bone from the *Iguanodon* forelimbs also failed to correspond in structure with the bone of modern reptiles. Oddly, it now occurred to him that the *Iguanodon* bone was more analogous to those of herbivorous *mammals*. Why should ancient reptiles bear any resemblance to modern mammals? Could this be used in any way to oppose the French progressionists?

While Owen was preparing his report on the giant reptiles, remarkable new evidence was discovered which, at first sight, appeared to support the progressionists. As Roderick Murchison, currently a senior member of the Geological Society, was pushing ahead with his research into the complex sequence of the Transition rocks, the history of life on the planet was moving into a new phase.

Murchison's early studies in Wales had shown that the ancient Silurian rocks had a highly characteristic fauna of trilobites, creatures with a segmented skeleton and multilensed eyes, and other marine invertebrates. During the 1830s, geologists in Europe found similar rocks in many locations in Europe, and gradually it became clear that the Silurian period was not of just local interest, but global. These ancient rocks held the first signs of life in primitive seas, and were as significant in the history of the world as the Age of Reptiles in the Secondary and the Age of Mammals in the Tertiary. In 1839, Murchison published *The Silurian System*, summarising his evidence.

Roderick Murchison believed that he had identified the earliest records of life, and so he was most concerned when Henry de la Beche, his colleague at the Geological Society, claimed that there were rocks in Devon which, he believed, were part of the Transition sequence and yet contained *land* plants such as algae, mosses and lichens. According to Murchison, this wasn't possible: there were no land plants before the Silurian.

In 1840, Murchison set out to resolve the classification of the Transition rocks, embarking on a continental tour which took him as far as Russia in a search for evidence. Eventually, he found sites where the order of these ancient rocks could be clarified. He showed that the Devonian rocks that Henry de la Beche had found with primitive land plants were formed at the same time as the rock known as 'Old Red Sandstone'. This contained the first vertebrates: fossil fish with strange armour, bony skulls and a thick shell-like covering. Both of these Devonian rocks lay *above* the Silurian with its marine invertebrates, and *below* the Carboniferous rocks, or coal measures, in which giant tropical forests had thrived.

Murchison's study, apart from disproving de la Beche's assertion, enabled him to define another layer of rock, or period in time: the Devonian, in which fishes first appeared in the fossil record. His work provided further evidence for a sequence, or progression, in the history of life. The marine invertebrates of the Silurian were followed by the fishes of the Devonian – the earliest vertebrate life.

But what happened in the deepest layers of the Silurian? asked his colleagues at the Geological Society. Could the very first signs of life be traced there? Did animal life just leap into existence in the fossil record somewhere during the Silurian period?

Others joined in the hunt, eager to claim the glory for tracking down 'the vestiges of Creation'. Travelling in central Wales, Professor Sedgewick of Cambridge University identified a 'Cambrian period' (called after 'Cambria', the old name for Wales) below the Silurian. This contained a 'primordial' fauna comparable to the marine forms Murchison described in the Silurian: molluscs such as shellfish, trilobites and brachiopods, shelled creatures similar to bivalves. Sedgewick hoped to be credited for uncovering the very first signs of life. To his surprise, invertebrates of some complexity such as trilobites seemed to appear from nowhere in the Cambrian rocks, but before this, life just petered out. While Sedgewick had pushed back the timing of the origins of Creation, the creation of life itself remained as inexplicable and mysterious as before.

In view of the accumulating evidence of changes in animal life over time, in 1841 John Phillips proposed a new way of classifying geological time. Phillips was the nephew of William Smith, who had pioneered some of the earliest studies of strata in England. His naming was still highly symbolic, as there was still no way of measuring time; the eras were periods of unmeasured distance, the earth's antiquity still unknown. Phillips suggested that the old divisions of Primary, Transition,

Secondary and Tertiary should be replaced with names that reflected the significance of the fossil evidence. The Primary rocks, with no traces of life, became the *Azoic* era, (from the Greek *zoe*, or 'life'). The Transition series became the *Palaeozoic* era, meaning 'ancient life'. The Secondary rock, which included the fossil reptiles, became the *Mesozoic* era, or 'middle life'. The Tertiary rocks became the *Cenozoic* era, 'newer forms of life'.

At the Geological Society, the consequence of the renaming of the eras and the accumulating evidence of fossil remains brought with startling clarity to an unwilling membership the idea of the progression of life over time. There were no fossils in the lowermost rocks of the Primary, or Azoic, era. In the Transition rocks, during the Palaeozoic era, ancient life began to appear. The lowest rocks in the sequence, from the Cambrian period, had the smallest variety of plant and animal life, such as marine trilobites. This was followed by the Silurian period, when invertebrates dominated the shallow seas. Then came the Devonian period, populated with bizarre, sometimes armoured fish, as well as land plants and corals. Above this were the coal measures, or Carboniferous period, where plants and large tropical forests flourished. This was followed by the next major era, the Mesozoic, corresponding to the old Secondary in which the globe experienced an Age of Reptiles as described by Mantell ten years earlier. Then came Cuvier's Age of Mammals in the Cenozoic era, at the end of which Man himself appeared.

A grand chronology of life was beginning to emerge. The gentlemen of the learned scientific societies were facing increasing difficulties in their attempts to explain away – without resorting to the claims of the evolutionists – the evidence that they themselves had gathered. Yet Buckland, Sedgewick and others were still united in their belief that the wonders of Nature – 'All creatures great and small' – were reflecting the glory of God. It was in this context that Owen's ideas on the giant fossil reptiles of the Mesozoic era were eagerly awaited.

In August 1841, the carriages of the scientific gentry descended on the West Country for the annual meeting of the BAAS, which was to be held in Plymouth. Mr and Mrs Owen travelled by boat from Southampton and stayed with an acquaintance, one Lieutenant-Colonel Hamilton-Smith. The President of the BAAS that year was also an old friend, the Reverend Professor William Whewell, who had attended the same school as Owen.

In his opening remarks, the Reverend Whewell announced with some pride that the speakers 'were the most gifted and eminent cultivators of science in the country'. He went on to promote the event: 'we have had experiments carried on at furnaces and in iron-works, on rail-roads and canals, in mines and harbours, with steam-engines and steam vessels, upon a scale which no Institution, however great, could hope to reach'.

Mr and Mrs Owen spent their first day visiting the geological section of the British Association, where they heard speeches on the latest geological finds from Professors Sedgewick and Buckland. On 2 August Richard Owen was invited to read his report. A very distinguished audience gathered, with Henry de la Beche in the chair. Owen rose: 'The present and concluding part of my *Report on British Fossil Reptiles* contains an account of the remains of the Crocodilian, Lacertian, Pterodactylian, Chelonian, Ophidian and Batrachian reptiles.' He proceeded to anatomise every single species of ancient reptile in a highly technical account, which lasted for two and half hours.

No sooner had he finished than Professor Buckland stepped forward and 'acknowledged Owen's labours, and the interest with which his report had been heard by the audience, in very complimentary terms'. With some pride, Owen told his sister shortly afterwards: 'My report gave such satisfaction that the Association immediately voted me 250 pounds for the expense of engraving the drawings and 250 pounds more for another report.'

But for Gideon Mantell, who had been unable to attend the meeting and so read an account of Owen's talk in the *Literary Gazette* on 14 August, it was devastating. He read the report with mounting dismay. Owen had classified the saurians, or lizards, into four divisions: firstly, the 'Enaliosauria' such as the ichthyosaurs and plesiosaurs, a group originally named by Conybeare, with typical lizard-like characteristics such as two openings in the skull. In the second division he had classified all the ancient crocodiles, 'the Crocodilian Sauria', many of which had been identified by Georges Cuvier. The next division included the ptero-dactyls, the flying lizards. Finally, under the category 'Lacertians', Owen grouped *Iguanodon*, *Hylaeosaurus* and *Megalosaurus* as 'very singular and very gigantic species which have now utterly perished'.

None of this was particularly controversial, and Owen was not the first to classify these three land reptiles together, since the German naturalist Hermann von Meyer had done so ten years earlier. However, as Mantell read the report, it became clear that Owen had used his plat-form to attack his rivals.

Firstly, condemning the early evolutionists, Owen pursued Buckland's argument outlined in the *Bridgewater Treatises*, claiming that the ancient reptiles were *superior* to reptiles of the present day. Among the reptiles, there was no progression in the fossil record from simple to complex beings. At one stroke, in this anatomist's hands the apparent progression of life was nothing more than a mirage.

'Owen's grand conclusion, so essential to science and our knowledge of creation,' enthused the *Gazette*, 'is that there was no graduation of one form into another . . . each were distinct instances of Creative Power, living proofs of a Divine will and the works of a Divine hand ever super-intending and ruling the existence of our world'. As though he had peered into the process of Creation itself, Owen had said: 'the evidence . . . permits of no other conclusion than that the different species of Reptiles were suddenly introduced upon the earth's surface'. Furthermore, from the very beginning they possessed characteristics that were 'originally impressed upon them at their Creation'. The giant reptiles were created by God as the most fitting forms of existence for the primitive earth, and thus they were firmly placed within the religious orthodoxy of early Victorian England.

Mantell had no quarrel with Owen's anti-evolutionary argument, but throughout the report ran a series of pointed slurs on Mantell's work. Owen ridiculed him for even attempting to seek similarities between ancient and modern reptiles, such as *Iguanodon* and the iguana: 'there is no existing lizard which offers such important differences in the structure of the teeth, the forms of the vertebrae, or of the long bones in comparison with the Iguana, as does the Iguanodon,' Owen declared. 'A very false notion would be entertained of that extinct Saurian, unless these differences were duly appreciated and allowed their full value.' In a pointed attack on Mantell, before the influential gentlemen of the audience, Owen claimed even the name '*Iguanodon*' was totally inappropriate.

Mantell read on: 'Professor Owen's examination of the numerous specimens now collected in the different localities enabled him to add many additional facts.' He described features observed by Mantell as though he himself was the first to recognise them, pointing up differences in interpretation as though he alone was correct. For example, fossils that Mantell conjectured belonged to *Iguanodon*'s forefoot Owen claimed were part of *Hylaeosaurus*; teeth of *Hylaeosaurus* and numerous other little details were also redefined, although the evidence was inconclusive.

For Mantell this was 'unworthy piracy and ingratitude', which was incomprehensible to him. Recognising that few in the audience would have had enough knowledge to appreciate how his ideas had been appropriated, he decided to set the record straight. He drafted a letter to the editor of the *Literary Gazette*, which was published on 28 August 1841. 'While expressing my admiration for the report,' he began, 'I beg permission to comment on a few statements which are not quite correct.'

Firstly, Mantell explained why he had adopted the name '*Iguanodon*': 'in my original memoir of 1825, it was distinctly stated that the name proposed . . . referred only to the general resemblance in external form of the fossil teeth, with those of the Iguana'. Sixteen years ago, Mantell said, there had been a shortage of evidence and no way of comparing the internal structure of the teeth with a microscope. Then, he pointed out, fossils identified by Owen in his report had already been described by himself, at an earlier date: 'the plano-concave vertebrae, the peculiar character of the femur and of the other bones of the extremities have been figured and described in my works'. As regards the unknown teeth that Owen attributed to *Hylaeosaurus*, 'in common fairness,' continued

Mantell, 'it should also have been mentioned that the first reference to the teeth in question to the Hylaeosaurus was made by me four years ago'. He pointed up similar faults in Owen's summaries of his, Mantell's, fossil turtles, but concluded, 'I beg most distinctly to disclaim any intention of attributing unfairness, either to the writer of the abstract or to the illustrious palaeontologist.'

This exchange of views within the pages of the *Literary Gazette* stoked the competitive spirit between the two men. During the autumn of 1841 Owen took every opportunity to improve and update his Plymouth speech before it was published eight months later. The British Association for the Advancement of Science did not normally allow extensive rewriting of talks prior to their publication. But for those within the inner circles of the BAAS, where Owen now found himself, it was possible to get away with redrafting, so long as it was not made public.

Meanwhile, to re-establish his supremacy in the field, backed by the Royal Society, Mantell was also planning another study: 'A Memoir on the Fossil Reptiles of South-East England'. This paper contained valuable new ideas; among other things, Mantell slightly reduced the size estimates for *Iguanodon*. More important, he was beginning to recognise that the forearms in *Iguanodon* were much smaller and more slender than the hind-legs, and might have been used for seizing vegetation rather than just for walking. This was to be a crucial factor in interpreting the anatomy and appearance of the animals correctly, which Owen had missed.

To advance his study, Mantell was keen to seek out any new fossils that might provide further insights into the ancient giant reptiles. Taking advantage of the railways, he went to quarries in Wiltshire and to his old haunts in the Weald in Sussex in search of more evidence. He was very pleased with a fossil he purchased from his acquaintance Mr Bensted in Maidstone, Kent, of the shell, ribs and vertebrae of an ancient turtle, and even permitted Richard Owen to come and see it in Clapham. Nevertheless, the antagonism continued to simmer between the two.

A few days after Owen's visit, during a routine visit to a patient on 11 October 1841, an event occurred that was to dramatically change

Mantell's life. The disaster happened in seconds. He was travelling by carriage along Clapham Common when the coachman lost control of the horses. Mantell tried to seize the tangled reins but was flung to the ground. The wheels grazed his head and he was dragged along the ground for some distance. In the violence of the fall he severely damaged his spine.

Mantell tried to recover at home, but gradually numbness spread from his foot. As the days passed, confined within his home in Clapham, he could no longer walk. The paralysis was spreading.

At Lincoln's Inn Fields, it is likely that Richard Owen was only too aware of his rival's misfortune. While Mantell could barely move, Owen continued to seek out 'every specimen available' that might shed further light on *Iguanodon*, *Hylaeosaurus* and *Megalosaurus*. Funded by the BAAS, he was immersed in redrafting his report for publication. Historian of science Professor Hugh Torrens has provided convincing evidence that Owen's key insights leading to the naming of the 'dinosaurs' as a distinct group occurred *after* his talk at the BAAS in August 1841, while he was writing up his report, and that it was almost certainly spurred on by his constant vying with Mantell. The more Owen studied this 'Lacertian' division, the more he recognised the remarkable characteristics that set them apart from the other divisions. Although Buckland and Mantell had portrayed these giant reptiles as lizard-like, using his expert anatomical skills Owen began to recognise that the giant thigh bone of *Megalosaurus* or *Iguanodon* was nothing like the curved femur of the crocodile.

The straight, vertical shaft of the femur was at *right angles* to the inward-turned head of the bone which fitted into the pelvis, like a mammal's. The implication was that these ancient creatures walked with their legs descending straight below the body like a mammal, not with their hind-legs sprawling to the side, like a lizard. Just like the bone seen in cross-section under the microscope, here was yet another mammalian characteristic. Thus the image of *Iguanodon* or *Megalosaurus* as giant crawling lizards, portrayed by Mantell for two decades, suddenly disintegrated. These were, for Owen, the finest reptiles that ever existed, almost as sophisticated as mammals in their structure, with long, hollow limb bones bearing prominent processes, or projections, for muscle attachment showing that they moved on land like mammals. As he contemplated their mammal-like features, Owen concluded that Mantell's estimates of the size of these beasts were grossly overestimated.

The question of size had been in Owen's mind since his ally, the collector George Holmes in Horsham, had come across more *Iguanodon* bones, even larger than those previously reported. The fossils were so large, even the bones of the claw were six times the equivalent bones in an elephant.

Mantell was too ill to travel to see the specimen. However, Holmes knew that Mantell calculated the size of the ancient beast by comparing each bone to the corresponding bone in the iguana, as he showed in his book *Geology of South-East England*. If Mantell's calculations were adopted, declared Holmes, 'comparing the proportionate size of the same bone in another of a similar genus . . . the largest ungual phalanx [bones of the fingers and toes] of the Great Horsham Iguanodon would give the animal a length of 200 feet'.

Richard Owen, alerted by Holmes, seized upon the new information with relish. The amount of muscle needed to lift such huge bones would weigh so heavily that such a beast could not possibly move. Mantell's calculations would produce an animal that was surely just too big to make biological sense. It was abundantly clear to Owen that a radically new method for estimating the size of the ancient reptiles was needed. Since he was becoming increasingly convinced that they were in many ways more comparable to modern mammals, rather than to modern lizards, he boldly came up with a new approach.

He measured the length of the vertebrae of the ancient reptiles, and then guessed their total number from head to toe taking account of the proportions of large pachydermal, or thick-skinned, mammals such as the elephant or the extinct *Megatherium*, rather than those of lizards. This led him to reduce the size of *Iguanodon* drastically. Given the very few caudal, or tail, vertebrae that had been discovered, Owen considered it 'very improbable' that the beast had as long a tail as the iguana; indeed, it was in all probability shorter even than the crocodile, around 13 feet. He estimated that there were twenty-four vertebrae each 5 inches long forming the trunk, and taking account of the sacrum, this gave a length of 12 feet for the main body frame. The head of *Iguanodon*, Owen speculated, was about 3 feet long.

So Mantell's one-hundred-foot beast shrank to a mere twenty-eight feet. Owen was so confident that his method was right, he proudly claimed that he needed only to study one fossil vertebra to 'give the length of the whole animal more correctly than any other plan hitherto adopted'.

At around the same time in the late autumn or winter of 1841, Owen had yet another key insight. A new bone of an *Iguanodon* had just been found in the Isle of Wight and purchased by William Saull in the City of London. It was the sacrum, or lower part of the spine, of the creature, the first that had been uncovered. Under normal circumstances, nothing would have stopped Mantell setting off to see a new *Iguanodon* bone. Yet despite weeks of rest, he was still distressed by 'numbness and paralysis' in his lower limbs and had 'great pain upon standing'. Consequently it was Owen, not Mantell, who hurried to Saull's collection in Aldersgate Street.

As Owen carefully measured the ancient fossil, it suddenly dawned on him that the *Iguanodon* sacrum had an identical characteristic to the sacrum of the *Megalosaurus* that Buckland had shown him, on display in Oxford's Ashmolean Museum for over two decades. The five sacral vertebrae forming the lower part of the spine in the *Megalosaurus* were *fused*. The newly discovered *Iguanodon* sacrum in Saull's museum, he noted with fascination, was fused in exactly the same way! *Megalosaurus* and *Iguanodon* – giant carnivore and herbivore – could be linked anatomically by this unique characteristic. He began to grasp the significance of this crucial feature: it was both beautifully simple and utterly compelling. A fused sacrum would confer tremendous strength to the backbone, enabling the giant reptiles to support their muscular tails and huge bodies. It was the perfect adaptation for living on land which the other three saurian divisions did not share. The sea lizards, the flying lizards and the amphibious crocodile division did not have a fused

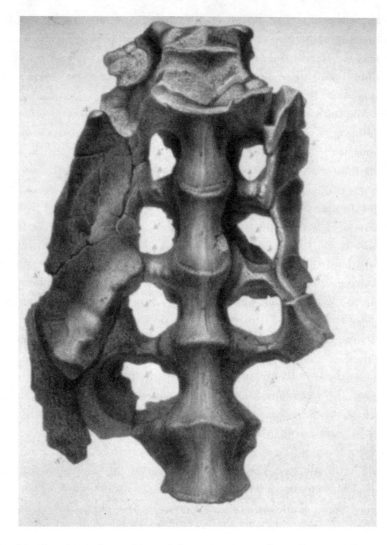

The fused sacral vertebrae of the *Iguanodon* – a key adaptation that
enabled the giant reptiles to live on land.

sacrum. Mammals, including humans, do have a fused sacrum, although it is fused in a different way.

Richard Owen began to realise that there were anatomical features that united the 'Lacertians' into a distinct group and fitted them superbly for living on land. Unlike the marine lizards or the pterodactyls that Mary Anning had helped to bring to light, the 'Lacertian' division had key defining characteristics. They were reptiles, and had scaly skin and laid eggs, but they possessed mammal-like characteristics in the shape and alignment of the limb bones and the sacrum. They did not sprawl like a crocodile, but moved on upright, pillar-like legs: these were reptiles designed for walking on land. They could be defined as a distinctive group of land-dwelling reptiles that walked with straight legs tucked up underneath their bodies. For Owen they embodied a form where the 'Reptilian type of structure made the nearest approach to Mammals'. He decided they needed a special name, in recognition.

Over the next few weeks he discussed possible names with geological friends and philologists. Keen to capture the characteristics that set these beasts apart from any that had ever existed, he seized upon the idea of using the Greek words *deinos*, meaning 'terrible' or 'fearfully great', and *sauros*, meaning 'lizard'. *Deinos*, a word used by Homer, also implies 'inconceivable', 'unknowable'.

Back in his study in the Royal College of Surgeons, he added these observations to his report of the previous August.

> The combination of such characters, some, as the sacral ones, altogether peculiar among Reptiles, others borrowed, as it were, from groups now distinct from each other, and all manifested by creatures far surpassing in size the largest of existing reptiles, will, it is presumed, be deemed sufficient ground for establishing a distinct tribe or suborder of Saurian Reptiles for which I would propose the name of *'Dinosauria'*.

In these few words, as he quietly redrafted his paper on that fateful afternoon, Richard Owen sealed the fate of Gideon Mantell. In this giant conceptual leap as he defined the characteristics of his '*Dinosauria*', he cast the spotlight on his brilliance at interpreting the fossil record. Although Mantell had known of the existence of fossil reptiles for years, in coining the term 'dinosaur', and presenting them as a distinct group of the most advanced reptiles that had ever lived, Owen was to receive the credit for their discovery. The disparate findings of the previous two decades suddenly crystalised into a unique form and an identity of their own.

Glorying in his new creation, he proclaimed: 'No reptile now exists which combines a complicated . . . dentition with limbs so proportionately large and strong, having such well-developed marrow bones, and sustaining the weight of the trunk by . . . so long and complicated a sacrum, as in the order *Dinosauria*.' Megalosaurs and *Iguanodons*, he said, 'rejoiced' in 'undeniably most perfect modifications of the Reptilian type. They attained the greatest bulk, and must have played the most conspicuous parts . . . as devourers of animals and vegetables, that this earth has ever witnessed . . . in cold-blooded animals.'

Striking a not uncharacteristic self-congratulatory note, he concluded: 'A too cautious observer would, perhaps have shrunk from such speculations . . . but the sincere and ardent searcher after truth, in exploring the dark regions of the past, must feel himself bound to speak of whatever a ray from the intellectual torch may reach, even though the features of that object should be but dimly revealed.'

Rays from his 'intellectual torch' they may well have been, but it may be, too, that Owen deliberately obscured the timing of his key insights. For when copies of his report were finally issued in April 1842, many were wrongly dated August 1841. Quite how this error arose is unclear, but it has provided fine fuel for conspiracy theorists who have suggested that Owen did this to create the impression of having achieved his insights at an earlier date, well ahead of his rivals.

He had, in fact, overlooked significant evidence on the dinosaurs while compiling his report. He had failed to recognise that some of the other giant lizards that he had assigned to different divisions, such as *Streptospondylus*, *Cetiosaurus*, *Thecodontosaurus* and *Poekilopleuron*, were in fact all dinosaurs. Although the data was available to him, he missed it and, in keeping with earlier work, grouped only *Iguanodon, Megalosaurus* and *Hylaeosaurus* together as dinosaurs. He also had no idea of the dinosaurs' true appearance, visualising them as sturdy, four-footed rhinocerine beasts, with grotesque, clumsy limbs.

News of the *Dinosauria* gradually leaked out beyond the narrow band of scientific pioneers. When Gideon Mantell and William Buckland had first discovered evidence of giant fossil reptiles in the 1820s, the nation-wide press had been very limited. By the 1840s, with the advent of the railways and advances in printing, there were several daily papers and news could be read across the whole country. In addition, the *Penny Magazine*, the *Penny Cyclopaedia* and the *Magazine of Natural History* covered scientific issues. These advances inevitably gave Owen's insights a much higher profile than the earlier discoveries had achieved.

Owen became firmly established as 'the English Cuvier'. A portrait of him was commissioned to complement the portrait of Cuvier in the gallery at Drayton Manor, the home of the Prime Minister, Sir Robert Peel. To add to his growing list of successes he was also introduced into royal circles at this time. Escorted by William Buckland, he attended an evening party held by Lord Northampton, a former President of the Royal Society, where he was introduced to Prince Albert. Later, in the spring of 1842, Owen was asked to receive Prince Albert and the King of Prussia at the Royal Society, alongside the Reverend William Conybeare.

The honours were now accumulating, and recognition also brought patronage. On 1 November 1842 Owen returned home to find a letter from Whitehall waiting for him. It was from the Prime Minister:

•

Sir,
It is my duty to offer advice to Her Majesty in respect to the appropriation of a public fund . . . in recognition and reward . . . of distinguished public service . . . I shall have great satisfaction in proposing to Her Majesty, with your consent, that an annual pension from H. M. civil list of 200 pounds be granted to you.

Your acquiescence in this proposal will not in the slightest degree fetter your independence . . . My object . . . is that the favour of the Crown may be most worthily bestowed . . . to encourage that devotion to science for which you are so eminently distinguished . . .

In spite of the late hour, Owen 'immediately put his boots on again' and 'sallied forth to our good friend, Justice Broderip'. William Broderip, a lawyer and long-standing ally who also lived in Lincoln's Inn Fields, was just going to bed, but he hurriedly pulled on his dressing-gown and helped Owen to draft a suitable reply. In a gesture of uncharacteristic exuberance on the part of the anatomist and the barrister (still in his night apparel), some sherry was ceremonially poured on the ground – an impromptu libation of thanks to God. Later, Owen and Buckland called on the Prime Minister personally to express their thanks. 'Dr Buckland maintained chiefly, Sir Robert listening like a clever man and occasionally making remarks.' This was followed by a visit from the Prime Minister to the museum at the Royal College. Records show that Sir Robert 'stayed more than two hours and was much gratified by his visit'.

While Richard Owen became a star in Victorian England, Gideon Mantell's private correspondence reveals that he came to brood more and more on what seemed to him to be an injustice. Night after night he had chiselled away at the fossils, sacrificing his marriage and his profes-sional practice to those momentous finds that had proved seminal to the whole field. Two out of the three dinosaurs, *Hylaeosaurus* and *Iguanodon*, which formed the basis of Owen's famous classification of dinosaurs, were Mantell's discoveries. The only new feature, Mantell felt, in the other man's interpretation of *Iguanodon*, was the analysis of the sacrum.

But in Owen's rewritten report of 1842, page after page described bones first found by Mantell and interpreted by him at an earlier date. And while giving Mantell little credit, he made much of his mistakes: 'it is very obvious that the exaggerated resemblances of the Iguanodon to the Iguana have misled the Palaeontologists who have hitherto published the results of their calculations of the size of the Iguanodon,' Owen wrote, pointing out the absurdity of Mantell's deductions that would produce an animal two hundred feet long in the Horsham specimen. To diminish his rival still further, he even cited *Iguanodon* in his summary as though it had been discovered by Cuvier, not Mantell.

Gideon Mantell confided in his American friend, Professor Silliman, 'I have to regret a want of honour and I may say justice, towards those but for whose labour and zeal he could never have obtained the materials for his own reputation . . . He altered names which I had imposed, and stated many inferences as if originating from himself when I had long since published the same . . . I do believe he would have altered the names Iguanodon and Hylaeosaurus had I not sent the letter of remonstrance to the *Literary Gazette*.' Yet Owen was the man idolised by the public and known even to Queen Victoria. 'His treatment of you . . . is unjust and dishonourable and merits exposure,' urged Silliman.

Months had passed since the accident, but Mantell was still an invalid. The paralysis in the lower half of his body lasted intermittently for twelve weeks; then sensation slowly returned, accompanied by excruciating back pain. As the months passed, intense pain came to dominate his life, depriving him of sleep for days on end and reducing his driving ambition to a humble desire merely to cope with living each day. 'Almost dead from pain and fatigue,' he wrote in his diary. To his alarm, a tumour 'of considerable size' began to appear on the left side of his spine.

12

The Arch-hater

There lives more faith in honest doubt,
Believe me, than in half the creeds.

Alfred, Lord Tennyson, *In Memoriam*

Richard Owen, 'the English Cuvier', was in his prime. At his home
town of Lancaster, a celebration was arranged in his honour in
September 1842. 'We walked in procession to the Town Hall, Mr
Whewell, the Mayor, the MP for the town and myself . . . cheered by
all the humbler folks . . . We sat down to a most princely banquet . . .
on three raised state-seats at the head.' Metamorphosed from the once
humble apprentice working in the local gaol, Owen had become indistinguishable from the gentry.

At the Royal College of Surgeons he was promoted to joint-
Conservator with Clift, sharing responsibility for the Hunterian Museum
with his former master. He undertook a major research project on
'Fossil Mammalia' for the British Association, completed his
'Odontography' on vertebrate teeth and was planning a summary of
'British Fossil Reptiles'. Caroline, and their only son William, bore his
dedication to his work with scarcely a complaint, apart from the odd
occasion when the smell from preserved animals in their home became
too much. 'The presence of an elephant's brain on the premises made
me keep all the windows open, especially as the weather is very mild,'
Caroline noted once when an elephant had died at the Zoological
Society. 'I got R to smoke cigars all round the house.'

Soon after creating the 'dinosaurs', Owen became still more famous when a remarkable prediction he had made came true. In 1839, he had been presented with a curious six-inch shaft of bone from an unknown creature from New Zealand. Observing the honeycomb matrix of the bone and its hollow structure, he had reasoned that this was from the limb of a bird, but because of its size, he had deduced that the bird was large and unable to fly. With great insight, he declared that a great flightless bird must once have existed.

Four years later, a missionary in New Zealand sent a hamper of fossils to Professor Buckland. It contained bones from a large flightless bird, exactly as predicted by Owen. 'Every word comes true to the letter,' Justice Broderip enthused to Buckland. 'This is another proof of the powers of our great physiological friend.' The giant feathered monster, which could attain a height of twelve feet, became known as a moa or *Dinornis*.

Owen's brilliant prediction was brought to the attention of Prince Albert. Buckland described the occasion to Owen: 'Sir Robert Peel and his Royal guest were astounded at the height of the Dinornis, "the very height of this library,"' Sir Robert had declared. Prince Albert wanted to see the moa bones for himself. 'No work of Owen's created so much excitement', according to one report. 'Society, headed by Prince Albert, hurried to inspect the huge remains . . . and to be introduced to the fortunate necromancer, at whose bidding a phantom procession of strange creatures had suddenly stepped out of the past, into the present.'

Owen's studies on the moa highlighted an intriguing observation. The flightless birds, such as the giant moa or the small, wingless kiwi, were found in New Zealand. South America was inhabited by mammals that were very different from any from anywhere else, both in the past, such as the extinct *Megatherium*, and in the present, with the related sloth and armadillo. Australia proved to be yet another distinct province, with extinct marsupials such as the *Diprotodon* and the present-day kangaroo and wombat. 'With extinct, as with existing Mammalia,' Owen wrote, 'particular forms were assigned to particular provinces.'

This made nonsense of the notion that all animals dispersed from one centre at Noah's Ark, and highlighted the puzzle of the origin of species.

The distribution strongly suggested that animals in the different 'provinces' had originated separately. So were there different centres of Creation?

With these thoughts in mind, Owen's reaction to a sensational book published the next year was muted. In *Vestiges of the Natural History of Creation*, the anonymous author set out evidence from the fossil record for the progression of life from simple to complex forms, showing the possibility of evolution without the hand of God. 'The simplest and most primitive type . . . gave birth to the type next above it,' he wrote, 'and so on to the very highest.' Although the author could not define the law governing development, he was in no doubt that such a law existed as surely as the law of gravitation. The shocking implication, spelled out in terms the layman could understand, was that Man himself could be the pinnacle of evolution and was not specially created by God.

Such was his anxiety at publishing this view explicitly that the author, a journalist, Robert Chambers, went to enormous lengths to conceal his identity. Among leaders of science there was outrage, even horror. According to the Reverend Sedgewick at Cambridge, 'the seductions of the author . . . poison the springs of joyous thought . . . he has annulled all distinction between physical and moral . . . in the new jargon of a degrading materialism'. If the book is true, said Sedgewick, 'religion is a lie; human law is a mass of folly; morality is moonshine; and man and woman are only better beasts'. It was essential to scotch the 'serpent coils of false philosophy'. Friends turned to Owen to write a damning review. 'A real man in armour is required,' Murchison urged his colleague.

But Owen was curiously reticent. During these years he was formulating his own ideas to account for the progression of the fossil record. Through studies on vertebrate anatomy he aimed to understand 'homologies', or 'equivalent parts', in the different animal groups. The foreleg of a lizard, the flipper of a seal, the wing of a bird and the arm of man were all homologous structures, connected to comparable parts of the body. Owen immersed himself in the vertebrate skeleton, seeking out more and more homologies. His aim was to identify the 'Ideal

Archetype', the common design or ground plan which, he believed, formed a blueprint for all vertebrates.

The concept of a blueprint, or archetype, for all vertebrate life became very significant for Owen. He believed it was 'the Divine idea' in the mind of the Creator as Nature was brought into being. From the archetype, he reasoned, God could foresee every possible form of vertebrate life: 'the Divine Mind which planned the Archetype, also foreknew all its modifications'. This, for Owen, proved that 'the knowledge of such a being as Man must have existed before Man appeared'. In other words, Man was planned and foreseen by God, and was not the result of some materialistic process. However, he admitted, 'to what secondary laws the orderly progression of such organic phenomena may have been committed, we are yet ignorant'.

His complex ideas, expressed very simply, could allow him to accept that there had been a progression of life over time, 'from the first embodiment of the Vertebrate idea' until Man himself, the pinnacle of Creation, came into existence. But the laws governing this progression were Divine laws, put in place by the Creator at the beginning. For Owen, God had not created each new species – He had created the laws which allowed them to form.

Owen's theory, providing a skilled synthesis of different threads of evidence, was of prime importance for Victorian biology. He followed his hero, Cuvier, in believing that the animal kingdom fell into four major divisions that were quite distinct. Within each division, studies on homologies by Geoffroy Saint-Hilaire and others were extended into the concept of the 'Ideal Archetype', the plan in the Creator's mind which allowed him to integrate natural history with the Christian faith. As propounder of such notions, Owen commanded enormous respect among his colleagues and leading figures of the day and was fast acquiring astonishing power for a scientist.

Exploiting his influential contacts, he aimed to expand his empire further. He wanted to unite all the collections of natural history in the British Museum, the Royal College and elsewhere under one roof, creating a national museum that would rival the Muséum National d'Histoire Naturelle in Paris. At the very least, he hoped to combine the fossil collections at the British Museum with the specimens he supervised at the Royal College. As he told one of the British Museum trustees, 'of all the Natural History departments in the museum, I believe this to be most out of place there'. Undoubtedly, he had his eye on Gideon Mantell's collection. How much more fitting, Richard Owen reasoned, that all these splendid fossil collections should be combined with those under his care at the College of Surgeons, so as to best illustrate 'the order and laws of Nature'.

Gideon Mantell, recovering slowly from his accident at home in Clapham, was acutely aware of Owen's success. 'I am still quite an invalid,' he told Professor Silliman in April 1842; 'I cannot stoop, or use any exertion without producing loss of sensation and power in my limbs.' Over a period of nine months he consulted many leading physicians: Liston, Brodie, Bright, Lawrence, Stanley, Coulson and others. It was thought the tumour on his lower spine was pressing on nerves, causing the intense pain and occasional paralysis.

Confrontation with Owen was out of the question. Publicly, he would not attack Owen's 'unwarrantable conduct'; rather the reverse – he even applauded his 'elaborate and masterly paper'. Privately, he confided to Professor Silliman, 'I am too ill to care one straw about worldly reputation . . . My feelings are so subdued by illness that I am more than ever anxious to live in charity with all men; and shall pass over these matters, at least till a more suitable opportunity offers.' The day of reckoning would have to be postponed.

Meanwhile, he struggled to maintain his Clapham practice but it became increasingly obvious that he would have to give up medicine to stand any hope of recovery. 'I have submitted to my fate and am negotiating for a successor,' he wrote in 1843. He could, however, still write while lying on the sofa, using a special desk he had made for his daughter Hannah, one that 'I contrived for my sweet child'. Following the success of *The Wonders of Geology*, Mantell embarked on another book, *Medals of Creation*.

He entreated his wife to return to him, but she did not, moving instead to Exeter with her housekeeper, Hannah Brooks. Apart from Reginald, his youngest son who was at college studying engineering, Mantell heard little from his children. He supported their travels but felt their absence keenly. Months passed without hearing from Walter. The lack of letters was noted in his diary: 'it is six months since I heard from Walter', or 'not heard from Walter since last September'. Occasionally, he caught news of his progress in the *New Zealand Gazette*.

The sense of isolation added to his suffering: 'I am in a very precarious state,' he confided to Professor Silliman, 'but I feel grateful for the blessings that I still have within my reach . . . and I can still hope on to the end.' His American friend never lost faith in him. 'There is no correspondent out of my own family to whom I write so frequently and so

Gideon Mantell.

long letters as to yourself,' wrote Silliman, 'because you tell me that they cheer you under your trials and I would cheerfully devote many hours in the year to that object.'

The early 1840s also heralded a difficult time in the life of Mary Anning, whose discoveries had laid the foundations for Owen's first report on the marine fossil reptiles. One local inhabitant, Nellie Waring, recorded her impressions of Mary Anning at this time: 'her little shop was scantily furnished and her own dress always of the very plainest. There was Mrs Anning, the Fossil woman's mother too, a very old lady in a mob cap and large white apron, who sometimes came with feeble steps into the shop to help us in our selection . . . the two were devoted to one another.'

But Mary's mother died in 1842. Soon after this, rumours began to spread that Mary had taken to drink. Gradually, it was realised that she was suffering from breast cancer; the most readily available pain relief was alcohol. Recollections of her at this stage contrast with the Mary Anning of earlier years. According to Nellie Waring, 'she was very thin and had . . . large eyes which seemed to me to have a kindly consideration for her little customers'. She was 'very timid, very unpretending and very patient . . . She would serve us with the sweetest temper . . . never finding us too troublesome as we turned over her trays of curiosities and concluded by spending a few pence only, and this we might do as often as we liked without causing offence.'

As news of Mary's illness reached the members of the Geological Society in London, William Buckland again tried to raise a subscription for her. Buckland's time was no longer concentrated on 'undergroundology'. He had been appointed Dean of Westminster in 1845, one of the most powerful positions in the Anglican hierarchy. As Dean, Buckland gradually eased himself out of the front line of geological research and became more involved in administration, restoring the school and the Abbey and organising sanitary reforms. Although he successfully organised a fund for Mary Anning, there was little else he could do to help.

Increasingly confined within her shop, she remained devoted to science. In her commonplace book, she copied out articles on the planets and geology along with 'Moral Maxims' and the poems of Byron. She also wrote down prayers for morning and evening. These expressed modest aims for each day: she should try to greet each day with gratitude, give thanks to God for her past life and even thank Him for her days of illness. In the words of Henry de la Beche, 'she bore with fortitude the progress of cancer on her breast, until she finally sank beneath its ravages on 9th March, 1847'.

She was buried in the Lyme churchyard by the sea, at the summit of the disintegrating Church Cliffs. At the Geological Society Henry de la Beche, now President, wrote a eulogy in her honour – most unusually, since she was not a Fellow. 'I cannot close this notice of our losses by death,' he said, 'without adverting to that of one, who though not placed among even the easier classes of society, but who had to earn her daily bread by her labour, yet contributed by her talents and her untiring researches, in no small degree to our knowledge of the great Enalio-Saurians and other forms of gigantic life entombed in the vicinity of Lyme Regis.'

Fellows contributed funds for a stained-glass window in her honour at the parish church at Lyme, showing Mary tending to the poor and healing the sick. 'This window is sacred to the memory of Mary Anning of this parish,' reads the inscription, 'in commemoration of her usefulness in furthering the science of geology, as also of her benevolence of heart and integrity of life.' In the words of a report in Charles Dickens's journal *All the Year Round*: 'the carpenter's daughter has won a name for herself, and deserved to win it'.

In 1846, Richard Owen's reputation came under critical scrutiny at a moment when yet more honours were being bestowed on him. In November he was nominated for the Royal Society's prestigious Royal Medal for his paper on the belemnite, the extinct mollusc distantly related to the squid and cuttlefish. Using his supporters at the Royal Society, Owen had arranged for Mantell's 1841 study of *Iguanodon* to be refused consideration. Curiously, Owen himself was chairing the meeting at the Royal Society when his own paper on the belemnite was recommended for the award. However, this piece of work was not quite

as original as it appeared. The little sea creature had already been described by an amateur, one Mr Chaning Pearce.

Chaning Pearce had come across the strange fossil during the building of the Great Western Railway. Its body was composed of fifty
chambers, and had an ink sac and ten arms with pairs of hooks and
suckers. In 1842, four years before Owen, Pearce's findings had been
read before the Geological Society and he had named the creature
Belemnoteuthis. Owen had been present at the meeting and had heard all
of Pearce's observations.

When Owen addressed the Royal Society in November 1846, he
made no reference to Chaning Pearce's earlier work. Ignoring the
previous study on the creature, he casually proposed a different name:
Belemnites owenii. Unfortunately for Owen, this name was based on an
erroneous assumption. He failed to realise, as Pearce had correctly
observed, that the fossil creature belonged to a new and previously
unrecognised genus which lacked an external solid 'guard', or shield,
typical of a belemnite, but could be identified by a brown coating
forming the outside surface.

Although Owen received the Royal Medal, his conduct in the affair
did not go entirely unnoticed. Edward Charlesworth, the editor of the
London Geological Journal, condemned his failure to acknowledge the
earlier work of Pearce: 'like cases are so common as to constitute an evil
of no slight magnitude in the progress of scientific research'. But
although Charlesworth continued to attack the Hunterian Professor,
Owen had risen so far that he seemed almost immune to criticism. Now
a frequent dinner guest at Drayton Manor, the home of Sir Robert Peel,
he used his powerful position to appeal directly to the Prime Minister for
a national museum of natural history. Such was his status in Victorian
society that he was invited within a few months to discuss his plans at
Downing Street.

While Owen had his eye ever more keenly fixed on uniting all the
famous fossils under his supervision, Gideon Mantell had started a
second collection of his own, in spite of being so incapacitated.
'Although unable to walk but a short distance, my mind is generally

vigorous,' he wrote optimistically. His Clapham practice was dwindling and he could not find a successor, yet obstinately he still insisted on making geological excursions. The 'Country of the Iguanodon' and its exotic flora and fauna had become, in his mind, a 'land of Promise'. Although, he wrote, 'I must be content to have obtained a distant glimpse of this Promised Land', he was still irresistibly drawn towards it and longed to complete his understanding of *Iguanodon*. 'I am not without hope that a tooth or two may be found attached to a fragment of jaw and have promised a rich reward to my men if they find such a specimen.'

His daughter Ellen occasionally returned to help him, drawing illustrations for his books and accompanying him on outings. *Medals of Creation*, published in 1844, was a success, running to a second edition. At every opportunity, Mantell continued to make contacts with local collectors, seeking out more original fossils to describe. Once he went with Ellen as far as Heyford in Northamptonshire, the ancient seat of the Mantells – 'Alas, now in the hands of strangers,' he wrote. There seemed little hope of winning back the family seat and the honours that would have gone with it.

Then he began to receive fossils from an unexpected source. He had not seen his son Walter for eight years, since the day in September 1839 when he had departed for New Zealand. During 1845 he became increasingly worried about his son: 'received a letter from Walter dated April; he is penniless and without any prospect of employment'. He sent out money and hoped Walter would come home. Then, to his great surprise, in July 1847 Mantell received a letter from him saying that he had come across interesting fossils and proposed to ship them to his father. Walter's box arrived from New Zealand just before Christmas. Opening it, Mantell saw that it contained more than eight hundred specimens – 'in fine preservation,' he noted with pride. Walter's collection, he thought, was the best that had ever reached Europe, containing many rarities, including the bones of a large, flightless bird, the moa or *Dinornis*.

In an ironic twist, Walter's discoveries served to confirm the brilliance of Owen's insights ten years earlier. There were many more parts of the skeleton: a perfect skull, where previously only portions of crania had been found; eggshells, jaws and other bones. Almost incredibly, given the hostility between them, Mantell invited Owen to his new home in Chester Square and gave him Walter's rare and precious bones. It is possible that he was hoping to restore cordial relations with such a potentially powerful ally, or perhaps he recognised his own lack of knowledge of the moa. 'As Professor Owen has made the subject peculiarly his own,' Mantell told a friend, 'I determined to forgo the pride and pleasure of describing these new acquisitions and allow him to have use of all the novelties my son has collected.'

Walter's discoveries provoked yet more interest in the flightless birds. Since the bones were not properly fossilised and there were persistent rumours from the Maoris that giant birds had been sighted, some believed they might not be extinct. Walter hoped to make his fortune tracking down the first living specimen. His new-found interest in science delighted his father and prompted much correspondence.

Within a few months, Walter was appointed Commissioner for the Purchase of Lands by the Governor of New Zealand. He intended, as he toured the middle island, to study the natural history of the islands, and was determined to trace the elusive birds. 'If there is a live Moa, my son will catch it,' Gideon Mantell told his friends proudly.

Meanwhile Reginald, his younger son, had returned from America and was working as an engineer with Mr Brunel, building the Great Western Railway. As he was supervising the works between Chippenham and Trowbridge, his team uncovered superb fossil belemnites. Reginald's fossils proved that Owen's applauded study of belemnites, which had earned the Royal Medal, was wrong. The disinterred belemnites proved, as the amateur Chaning Pearce had maintained, that Owen had falsely ascribed features to the squid-like creature that it did not have. This belonged to that same distinct genus that Pearce had discovered: *Belemnoteuthis*.

Mantell, now armed with the evidence, could not resist taking on Richard Owen. He prepared a paper for the Royal Society describing the intricate details of the *Belemnoteuthis* anatomy and its shelly exterior

coating, or capsule. It was a minor detail in the interpretation of invertebrate anatomy, but a major setback for Owen. 'He was not one to admit
having been mistaken with good feeling,' Mantell wrote to Professor
Silliman. Almost absurdly, when the real fight was about the dinosaurs,
the battle lines became drawn over this small squid-like creature.

An unusually large number gathered to hear Gideon Mantell's paper
to the Royal Society in 1848. Although couched in the restrained
language of science, he felt that his comments made it clear that Richard
Owen's paper 'was a tissue of blunders from beginning to end'. However, Owen had ensured that he had supporters present. 'After the paper
was read, Professor Owen got up and made the most ungentlemanly and
uncalled for attack upon it,' wrote Mantell. 'He said that I ought not to
have presumed to occupy the time of the Royal Society . . . and after
ridiculing for half an hour all that I had written, sat down and was
actually applauded by many.' This prompted his old ally the Dean of
Westminster to rise and strongly defend Mantell's paper as 'in the
highest degree important'. According to the editor of the *London
Geological Journal*, Edward Charlesworth: 'there was a most animated
discussion in which all who took part, including Buckland, Bowerbank
and others, made a resolute stand against Owen on behalf of poor
Chaning's Genus, *Belemnoteuthis*'.

The point had been made. The alleged 'Newton of Natural History'
was not infallible. But as Mantell was well aware, Owen was intolerant
and resented 'that anyone put a foot upon the lowest step of his throne'.
It wasn't long before they were to clash again, this time over a fossil that
Mantell had dreamed of finding for years.

In March 1848, Gideon Mantell received an unexpected package from
a stranger, a Captain Lambart Brickenden. The Captain, who was the
proprietor of the quarries in the Tilgate Forest in Sussex, had uncovered
an *Iguanodon* jaw. It was not complete, but merely a part of the lower
jaw, over twenty inches long, very heavy and a rich umber colour. There
were sockets for seventeen or eighteen identically shaped teeth and two
tiny replacement teeth. Although the upper jaw was missing and the
adult teeth were no longer attached, the replacement teeth proved

the animal was reptilian. Here was the elusive evidence that Cuvier had urged Mantell to find in the early days, when no one believed that he had found a new reptile. And now the treasure he had looked for for so long and desired so desperately was found. Mantell was in no doubt of its significance: 'Here, after thirty years' search,' he wrote, 'is an unequivocal portion of the dental organs of that marvellous reptile.'

At the time there was a Chartist riot going on in London. Cannon were in place at the palace, and soldiers in the streets. Mantell waited until the crisis was past before he ventured out to the British Museum to compare the fossil with the jaws of other animals. To ease the burden, since he was in some pain, he collaborated with a skilled anatomist, Dr Alexander Melville, Professor of Zoology at Queen's College. Comparing the lower jaw with others in the museum, they wrote, 'we are at once struck with their remarkable deviation from all known types in the class of reptiles'.

The *Iguanodon* jaw had a curious combination of characteristics. Unlike living iguanas or the large extinct lizards whose jaws 'are armed with teeth to the anterior extremity', this jaw expanded at the front into a 'scoop-shaped projection' similar to the extended lower jaw of the mammalian sloth. From its teeth, they knew that *Iguanodon* masticated its food like modern ruminants, while the method of implantation of the teeth and their replacement cycle was more like those of reptiles.

Mantell was invited to deliver a paper on the jaw at the Royal Society on 18 May 1848. 'Although several hundred teeth . . . of Saurians have been collected,' he began, 'but a few fragments of the jaw have been discovered . . . It is therefore most gratifying to have it in my power to lay before the Royal Society . . . the first indisputable portion of the jaw of the Iguanodon hitherto brought to light.'

The replacement cycle of *Iguanodon* teeth, which he had longed to prove as a young man, he now described in detail. The formative pulp was in a distinct cavity on the inner side of the root of the tooth that it was destined to supplant. 'In the Iguanodon, the old teeth were retained until . . . the crown of the tooth, from abrasion by use above and removal of the fang by absorption below, was reduced to a mere disk, before it was finally shed.' Since, he reasoned, all fangs showed some sign of absorption, 'the formation of successional teeth was in constant progress at all periods of the animal's existence, as is the case in most of the Saurian reptiles'.

Cautiously, he attempted to calculate the size of the dinosaur's head. Since comparisons to the lower jaw of lizards suggested this bone represented nearly half the jaw, he estimated the total length of the jaw could be four feet. This, he acknowledged, was in disagreement with Professor Owen, who had claimed the largest *Iguanodon* head was only two and half feet long. To make his calculation, Owen had measured the length of six dorsal vertebrae, which in the iguana is equal to that of the lower jaw. 'But even if we take the short blunt-headed lizards as the scale, for example the Chameleons,' said Mantell, 'the length of jaw of this Iguanodon must have exceeded three feet.' There was, in fact, no way of proving the size of the head from the portion of jaw. Both were speculating, drawing their conclusions from analogies to different bones.

Mantell even attempted to define the soft tissues of the dinosaur's face and the muscular adaptations that would be needed to chew tough plants and leaves. Because of the large number of holes for blood vessels at the front of the jaw, he reasoned that these supplied the muscles and soft parts around the mouth. From this, he inferred that 'the under-lip was capable of being protruded and retracted' and together with a 'large, fleshy, prehensile tongue . . . formed a powerful instrument for seizing and cropping the leaves and branches'.

These ideas anticipated many later studies on the soft tissues of *Iguanodon* cheeks and mouth, although it is now known that it did not have a large, prehensile tongue. From the quantities of vegetables that it had to eat, 'there must have been a large development of the abdominal region,' Mantell reasoned. The rear and hind-legs he saw as bulky, 'presenting the unwieldy contour of those of the Hippopotamus and Rhinoceros, with horned toes, and bulky muscular legs'. The teeth and jaw, he concluded, 'demonstrate its power of mastication and the nature of its food'.

Finally, Mantell took a bold step and announced the existence of yet another dinosaur. An earlier discovery of a small fragment of lower jaw, he claimed, had been wrongly attributed to *Iguanodon*. He had examined fragments of bone and teeth from this unknown jaw under the microscope: it bore no resemblance to 'the very fine, dense tooth ivory of the Hylaeosaurus'. Nor did it match any other known dinosaur. Although more similar to *Iguanodon* than to anything else, it was not identical. Mantell concluded that it was in fact from a new dinosaur, which he named the 'Regnosaurus'.

But when Mantell had finished his talk, to his great dismay Professor Owen ceremoniously announced to the learned audience at the Royal Society that 'a smaller and more perfect specimen of the jaw had already been found at Horsham'. Owen aimed to show that some of Mantell's inferences were incorrect and that his *Iguanodon* jaw was not the first, as he had claimed. Mantell was astounded. He had no knowledge of any other fossil jaw of *Iguanodon*. Yet here was the professor, apparently eclipsing him once more.

It soon transpired that George Holmes, the Sussex collector patronised by Owen, had recently found a smaller specimen of part of the jaw from a young *Iguanodon*. Mantell's friend Captain Brickenden, who owned the Tilgate quarries, went to see Holmes and made drawings of his fossil for Mantell. Although Holmes's specimen showed more detail, Captain Brickenden was able to reassure Mantell that the second specimen affected none of his inferences. This incident, however, was one of several in which Holmes inadvertently stoked up the rivalry between Mantell and Owen.

Holmes acted on Owen's behalf, sometimes making it difficult for Mantell to see his collection in Horsham or even to make drawings. He kept Owen informed of Mantell's plans: 'the Doctor [Mantell] expressed his intention to coming down before long to see my Collection,' he advised Owen. 'I do hope that thou wilt not be behindhand with him in thy visit, if thou canst possibly make it convenient to come.' In another letter, Holmes told Owen that Mantell was researching the backbones of the giant reptiles. He described vertebrae that Mantell had identified and even pointed out an error that he might have made, which

Owen could verify by checking with the *Hylaeosaurus* in the British Museum. It is no surprise that Mantell came to view Holmes as 'a sly quaker' and a 'spy' for Owen.

By now, Mantell was receiving fossils from several different sources. Captain Brickenden continued to send consignments from the Sussex Weald, and Mantell was also in contact with collectors on the Isle of Wight, even fishermen in Brook Bay and Sandown Bay. He hoped to obtain enough vertebrae to reconstruct the spine. This would enable him to prove the overall length of *Iguanodon*, the size and mobility of the neck and tail; even the creature's bulk could be estimated by the way the ribs were attached and the size of the lumbar vertebrae.

He was becoming increasingly suspicious that vertebrae that Owen had attributed to different reptiles were, in fact, different parts of the spine of *Iguanodon*. Since he was so weak, he collaborated once more with Alexander Melville at the British Museum, who had the anatomical skills to help him take on Richard Owen. Mantell had been told the tumour was fast growing on his spine, at the site of the injury from the carriage accident, and there was no treatment available. His attempts to control the pain were becoming desperate. 'Took hot brandy, and water with brandy and laudanum, ether and camphor, hot air bath, inspired chloroform. All unavailing,' he wrote after one attack. Increasingly he turned to opiates for pain relief – at first laudanum in pharmaceutical doses and then liquor opii sedativus, an opium derivative. On one anniversary of Hannah's death he was too ill to visit her grave. Summoning all the resources that were left in his increasingly frail body, he was determined to get as far as was humanly possible to elucidate the true appearance of *Iguanodon*. But he knew he was running out of time and longed to be free of pain: 'living in the hope that death may give the imprisoned spirit freedom'.

In July 1848 Mantell read in *The Gentleman's Magazine* of the suicide of his former Curator in Brighton, George Richardson. Shocking details of Richardson's plight emerged at the inquest. He had had difficulties living within his curator's pay of less than £100 a year. As his debts had mounted, he had faced bankruptcy and feared the disgrace. Richardson had been found 'with his head nearly severed from his body, with a razor which lay near him'. Further investigations proved 'that the deceased had deliberately sat before the looking glass and cut his throat. The glass, chair and razor were covered in blood.' Gideon Mantell was shaken. Richardson had been a close ally in Brighton; his head, severed 'from ear to ear', was a gruesome image with which to end that episode in their lives. 'I deeply deplore this melancholy event. It has haunted me ever since I heard of it,' wrote Mantell.

As soon as Owen heard through Holmes of Mantell's interest in the *Iguanodon* backbone, he wrote to warn his rival against publication. He was wasting his time, he threatened Mantell, since his own study was now virtually complete: 'the first part of my work will appear soon after Xmas; 20 plates are already struck off, it will include the Reptiles of the Eocene formation. I shall next proceed to chalk, greensand and Wealden.' Even though Mantell was the original discoverer of *Iguanodon*, Owen was becoming increasingly territorial about it and eager to be first with any new insights on the beast.

Owen's actions merely served to goad Mantell into still greater effort. Almost absurdly, given his physical weakness, he and Melville devoted so many hours to their study that it was completed within a month of receiving Owen's warning. 'Dr Melville spent all day here, and while visiting my patients, he went on with his description of the vertebrae,' Mantell wrote on 15 January 1849. 'At eleven at night we finished the Memoir at last. Never was I so tired of a task of this nature before.'

Two months later, Gideon Mantell read his paper to the Royal Society. He did not have a complete backbone and acknowledged that 'there is still no clue to guide us through the labyrinth but analogy'. Nonetheless, he had made progress. When he had begun his studies thirty years earlier, large vertebrae of dissimilar forms had been 'vaguely assigned to the Iguanodon'. Many of these had subsequently been re-identified by Owen as part of different reptiles, such as *Streptospondylus* or *Cetiosaurus*. However, Mantell had acquired vertebrae from the Isle of Wight found together and showing such a 'close affinity of the bones . . . as to leave little doubt that they belong to the same animal'. Armed with this evidence, he demolished Owen's earlier identification of the vertebrae and showed they were in fact different parts of *Iguanodon*'s spine.

Professor Owen's *Streptospondylus* vertebrae became *Iguanodon*'s cervical, or neck, vertebrae, and his *Cetiosaurus* vertebrae became the posterior dorsals and caudal, or tail, spines of *Iguanodon*. Mantell also described and provided measurements of the other parts of the spine, the anterior dorsal and lumbar vertebrae. This was the first time the different types of vertebrae had been correctly identified. His analysis revealed that the *Iguanodon* vertebrae were wide and tall, capable of supporting a bulky frame. Although not widely appreciated at the time, recent analysis by Dr David Norman at the University of Cambridge 'proves Mantell and Melville to have been wholly correct in their conclusions'.

Most important of all, Gideon Mantell was the first to observe the small size of the humerus, the upper bone of the forelimb, which had major implications for recognising the true shape of the beast. As yet, no one had been able to identify the forelimbs with any certainty. Even though the upper forearm was embedded in the Maidstone specimen found in 1834, for years it had been overlooked since it was thought that these reptiles were four-footed creatures, with both fore- and hind-limbs of the same proportions. In the Maidstone *Iguanodon* the femur of the hind-leg was 33 inches long. There was no corresponding forelimb bone of equal size. Consequently, the humerus, which was only 20 inches long, was mistaken for a radius, the bone of the lower arm. Later, Owen suggested it might be a foot bone .

'The question however, is now decided,' announced Mantell, 'by the discovery of a bone found in the Wealden strata of the Isle of Wight associated with other remains of the Iguanodon and which is undoubtedly a humerus because it cannot possibly be referred to any other part of the skeleton.' It was also identical in shape to the bone that he had long suspected to be the humerus of the Maidstone specimen. In the Isle of Wight fossil the femur of the hind-leg was 4 feet 8 inches, and the humerus was 3 feet 2 inches. Allowing for compression, in both the Maidstone and Isle of Wight fossils, the humerus of the forelimb was a third smaller than the corresponding bone of the hind-limb.

This confirmed Mantell's far-sighted view that the forelimbs of *Iguanodon* were much less bulky than the hind-legs. They were 'long and slender and served as prehensile instruments . . . adapted for seizing and pulling down plants and branches of trees'. Unlike the hinder limbs and feet which were 'strong and massive as in the hippopotamus', to support its enormous carcass, the arms were capable of grabbing the lush tropical vegetation of ferns, cycads, reeds and conifers. *Iguanodon*, Mantell announced, was 'one of the most remarkable herbivorous terrestrial quadrupeds that ever trod the surface of our planet'. With some satisfaction, he went on: 'After a lapse of more than a quarter of a century, I conclude my attempts to restore the skeleton of the gigantic Saurian, of whose former existence a few isolated and water-worn teeth were the sole known indications.' With the exception of the bones of the skull, the sternum and the lower forearm, 'the entire skeleton may now be considered as determined'. The creature that had for so long occupied his mind was beginning to take shape.

Mantell was now challenging Owen's supremacy in the field of dinosaurs. While Owen had relished cutting Mantell's dinosaurs down to a mere 30 feet, there was accumulating evidence that some dinosaurs did indeed attain stupendous proportions. In the autumn of 1849 Mantell received the head of a tibia, or leg bone, of an Iguanodon that was a massive 58 inches in circumference. Soon friends told him of a miller at Malling Hill, near Lewes, who had uncovered another monstrous leg bone in the Weald. The new fossil proved to be 'a glorious specimen of a humerus'. At four and half feet it was the longest portion of arm bone yet found. Gideon Mantell noted with interest that 'it has not all the characters of a humerus of the Iguanodon'. But if it was not from an *Iguanodon*, then what was it?

This was not easy to resolve. After careful comparisons, he thought the bone bore most resemblance to lumbar vertebrae retrieved from the same pit and identified as part of the lizard known as '*Cetiosaurus*' by Owen. Mantell made a trip 'by express train' to Oxford to Buckland's museum, where there were other *Cetiosaurus* bones. These were very distinctive, with a spongy texture like the bones of a whale (hence the name *Cetiosaurus*, or 'whale-lizard'). But the new humerus did not quite match any of these bones. In fact, it was unlike any of the saurian leg bones Mantell had seen before.

There was only one other possible conclusion: that it was from an entirely new kind of dinosaur, perhaps larger than any yet discovered. In a pointed gesture directed at Professor Owen, Mantell proposed the name '*Colossosaurus*'.

Soon he was at work on a paper for the Royal Society on his new 'pet lizard', as it was endearingly nicknamed by Silliman in the *American Journal of Science*. He purchased the large humerus for around £8 and commissioned an artist to draw the specimen. By November 1849, he had decided on a slightly more subtle name for his new beast: '*Pelorosaurus*', from the Greek word *pelor*, or monster. Mantell's *Pelorosaurus* was the first named dinosaur in a family known today as the sauropods (meaning 'lizard-foot'), recognised as the largest creatures to have ever walked the earth.

A few months later, Gideon Mantell went on to identify yet another sauropod and his sixth dinosaur: *Cetiosaurus*. Owen had failed to recognise that *Cetiosaurus* was a dinosaur, and envisaged that it was related to crocodiles, 'strictly aquatic, probably of marine habits'. From specimens obtained on the Isle of Wight, Mantell could see the creature had a fused sacrum of the 'dinosaurian type'. The massive fused sacrum was one of Owen's defining characteristics of a dinosaur. Since Mantell saw the *Cetiosaurus* sacrum first, he was able to beat Owen with his own definitions and correctly identify the animal as a dinosaur.

The giant sauropods are now known to include such dinosaurs as *Diplodocus*, *Brontosaurus* and *Apatosaurus*. They are characterised by very long necks and tails, and large bodies on pillar-like legs. Of all the dinosaurs, their skulls are smallest in relation to their body size. Using their long necks they could reach branches and leaves inaccessible to other dinosaurs, to consume vegetation. *Seismosaurus*, the largest dinosaur ever discovered, could exceed 120 feet. Mantell correctly estimated, just from the humerus, that *Pelorosaurus* could attain 80 feet.

With this success, towards the end of 1849 Gideon Mantell's name was proposed once more for the prestigious Royal Medal of the Royal Society. But he learned that the committee passed over his paper on *Iguanodon* because of Owen's disparaging remarks. On no less than three occasions the committee and Council of the Royal Society had met to decide the matter. Each time, Richard Owen did everything in his power to prevent the award being made to Mantell. 'All Mantell had done,' he argued, 'was collect the fossils and let others work them out!' Hearing of these closed sessions from a friend who was on the committee, Mantell was enraged. 'What a pity a man of so much talent should be so dastardly and envious,' he wrote to Professor Silliman. 'Professor Owen claimed my papers in the Transactions were unworthy for such an honour. Although he received it himself for his paper on the Belemnite, which has proved to be utterly erroneous!'

Spurred on by his supporters, Mantell sent his paper on *Iguanodon* to the Royal Society and asked the Council to reconsider the award of the Royal Medal. In his own mind, it was the significance of his life's work that was being debated. The years and years of being slighted by Owen finally became too much to bear. At stake here was who would get the credit for interpreting the key dinosaur fossils and defining the ancient creatures. But Owen was not prepared to acknowledge that his own work on dinosaurs was built on foundations laid by Mantell.

Under the watchful eye of the scientific community, justice had to be seen to be done. A fourth meeting of the Council and committee was called. Once more, Owen launched into an attack on Mantell, ridiculing his speculation that *Iguanodon* had cheeks and soft parts covering its gums, and pouring scorn on his work. This time Sir Charles Lyell was present, and rallied to his old friend's defence. He discussed the merits of Mantell's studies, pointed out how often Owen had used Mantell's research in his own work, and quoted the high praise Mantell had received from Cuvier. Professor Buckland, too, had written to the committee of the Royal Society stating that all of Mantell's papers, on *Iguanodon*, the foraminifera (marine organisms with perforated shells) and – in a pointed attack on Owen – the belemnites, qualified for the highest honours the Society could bestow. Consequently, on 30 November 1849, the Royal Medal *was* awarded to Mantell. Only Owen and one other member of the council cast a vote against him. When Mantell was invited into the meeting-room and informed of the decision, he noticed that 'Owen sat opposite me and looked the picture of malevolence'.

Later, when Mantell was at a meeting at the Royal Society, Owen came up and shook hands with people near him, then stretched out his hand to Mantell, saying what a great pleasure it was to see him there. Was this, on Owen's part, just a trivial and meaningless gesture, or was it perhaps a signal of reconciliation? Mantell, knowing full well how Owen had acted on the Royal Society Council, saw his duplicitous hand-shake, bowed and declined it. He would not touch the hand of the one who had so effortlessly tried to take it all away from him. Years of disappointment and frustration became crystalised in his seething irritation at the younger man, who had collected all the glory for the dinosaurs. To his mind he had been treated cheaply.

Despite Mantell's acknowledged scholarship, Richard Owen never relaxed his grip on his rival. It was beyond him to concede any ground to the opposition. Possibly in an effort to pre-empt gossip about his own Royal Medal, under his influence, a glowing account of his analysis of the belemnite appeared later in the *Quarterly Review*. Owen's friend, Justice Broderip, had written the article, with the text apparently checked and corrected by Owen himself.

As the vendetta escalated, Owen took yet more steps to undermine Mantell. Owen was hoping to publish a definitive work on British fossil reptiles. In October 1850, he applied to the Council of the Royal Society for permission to take many impressions from illustrations of fossil reptiles published in the Society's journal. However, Owen failed to mention to the Council that some of these were, in fact, Mantell's carefully researched illustrations. Instead, he implied that this was all his work, stating that they were 'plates described by him in his 1842 report on *British Fossil Reptiles*'. Even Captain Brickenden's famous *Iguanodon* jaw, which meant so much to Mantell, Owen implied was one of his own. Consequently, at a meeting of the Council on 24 October 1850, his request was granted.

When Mantell heard what had happened he could not contain his fury. He discussed the matter with Sir Charles Lyell, who 'expressed his astonishment at such conduct,' Mantell wrote. He confided in Captain Brickenden: 'you cannot imagine the annoyance I have had again from Professor Owen; he is not satisfied with monopolising everything he can from my first discovered rocks but tries to rob me of the few things that I got from friends . . . He is more jealous and envious than ever!'

A month later, a special Council meeting was convened in the grand committee room of the Royal Society. Gideon Mantell had conclusive proof that some of the plates were not Owen's from 1841, because many of the fossils had been found after this date. Owen was forced to backtrack; for him it was a trivial matter – he uttered an effusive, meaningless apology. But for Mantell: 'every one of the Members of the council present seemed to be convinced that Owen, for once, had been caught and exposed in his duplicity'. Owen, he thought, was 'overpaid, over-praised, and cursed with a jealous monopolising spirit!'

For so long contemptuous of anyone who stepped across his path, Richard Owen was fast becoming a law unto himself. He clashed with Charles Lyell, which some believed was retaliation for Lyell not supporting Owen's manoeuvres at the British Museum, and with Alexander Melville, a supporter of Mantell's. The affable biologist Hugh Falconer also fell victim to Owen's schemes. Their dispute began when Owen 'stole' the naming of an American elephant from Falconer; hostilities later degenerated into bitter personal attacks, as they argued over the characteristics of certain extinct marsupials. Owen even went so far as to pick a fight with the Queen's dentist, Alexander Nasmyth. Owen claimed authorship of Nasmyth's interpretations of the structure and growth of teeth, and accused Nasmyth of plagiarising the ideas of others, exactly as he himself had done with the 'dinosaur'. Eventually, loyal allies, too, such as George Holmes, were to turn against Owen. Holmes complained that he was being 'shamefully treated' when he discovered that some of his fossils were not being returned as agreed, but appropri-

Thomas Henry Huxley.

ated into Owen's collection at the Royal College. Holmes may have realised that he had become merely one of Owen's weapons in the battle against Mantell.

Owen seemed to thrive on feuds and antagonism, wounding his rivals with almost the same clinical satisfaction with which he tackled his dissections. He was described in one biography as a 'social experimenter with a penchant for sadism and mystification', and in another as 'addicted to acrimonious controversy' and driven by arrogance and jealousy. His rancorous disputes were to make a deep impression on one young anatomist who was struggling to establish himself in London – Thomas Henry Huxley.

Huxley had returned from a voyage on HMS *Rattlesnake* in 1850, then rapidly made a name for himself with studies on marine invertebrates that he had observed on his travels to Australia and New Guinea. Although these earned him a Royal Medal at the young age of twenty-eight, he struggled for years to obtain a paid scientific post: 'a man of science may earn great distinction, but not bread,' he told a friend. Huxley turned to Owen for help, but then became suspicious that Owen was not doing enough, although he did in fact write the young man a number of references. 'It is astonishing with what an intense hatred Owen is regarded by most of his contemporaries, with Mantell as arch-hater,' Huxley was to observe.

As if innocently oblivious of the mayhem he created in scientific circles, Owen portrayed Man in his public lectures as the pinnacle of Creation, made for God's purpose. 'The supreme work of Creation has been accomplished . . . for the service of the Soul. Think what it may become – the Temple of the Holy Spirit!' he declared with galling virtue. 'Defile it not. Seek rather to adorn it with . . . that fair furniture, moral and intellectual, which it is your inestimable privilege to acquire.' With his 'Archetypes' and 'Divine Plans' and other technicalities which added not a little to the obscurity of his subject, no one, it seemed, could check his upward rise.

In March 1850, Owen was summoned to his first levee, or formal reception, at the palace, where he was presented to the Prince Consort by the Earl of Carlisle. In the same month, he was appointed to the Prince's Council at Buckingham Palace to advise on the planned Great Exhibition. He immediately summoned a court tailor. With Caroline's help, he 'devised a very handsome and elegant attire, I think quite as good as any Court dress I saw. A rich sort of dahlia-brown cloth, with bright steel buttons, buckles, sword &c and white satin waistcoat with rich flowers embroidered. Lace cravat full and long, and the same for the cuffs. Cut steel loop in the cocked hat. All very fine, as Pepys would say.' Soon Owen was invited to tutor the children of the royal family.

While Owen lightly shrugged off any opposition, the feud between the two men was exacting an ever more terrible toll on Mantell. Gradually, he was reduced to a shadow of his former self. When he met Lyell on one occasion, his old friend seemed quite shocked at the decline in his physical appearance. The years of pain showed. One doctor held out promise of an operation on his spine, but it was decided that it was too risky. Eventually, his damaged spine was forced into a painful curvature and his whole body became horribly twisted, as if to emphasise the disappointments and frustrations that filled his mind.

Aware of the relentless progress of his illness, he continued to find a welcome relief in the letters from his son. Walter did not find a live moa, but he sent many fossils from New Zealand, including 'a matchless collection of fossil birds comprising nearly five hundred specimens'. One day in 1850, Mantell received a letter from him written from a mud hut, surrounded by six feet of snow, on a bleak shore of Bank's Peninsula. Walter was excited because he had heard that two hundred miles away were some large caverns with bones of unknown animals on stalagmite floors. He must investigate them. He had completed his work as Commissioner for the Purchase of Lands, and could now afford to buy a house and land and settle down as a farmer. He urged his father to abandon his struggles in London, and come and stay with him. Mantell did not join his son, but he made arrangements to send out six hundred samples of rocks and minerals, a fifty-drawer cabinet and all the equipment necessary to investigate the caves. His son would want for nothing in the pursuit of science.

13

Dinomania

Nor love thy life, nor hate; but what thou liv'st
Live well,
the rest leave then to Heaven.

> Milton, *Paradise Lost*, Book I, as quoted in
> Mantell's correspondence to Silliman

May the 1st, 1851: the official opening of the Great Exhibition. Joseph Paxton's glittering Crystal Palace covered nearly twenty acres of Hyde Park. Hotels and boarding-houses were fully booked up to twenty-five miles outside the capital. One thousand state carriages had arrived by midday, and two thousand cabs – every possible vehicle was put into service, all helping to make a London traffic jam that was to last for six months. Half a million people gathered in Hyde Park that day. Army troops were ready to fire a salute to Queen Victoria, a plan that caused concern. 'Thousands of ladies will be cut into mincemeat,' warned the *Times*, fearing that all the glass would shatter as the guns fired.

Inside the Crystal Palace – an enlarged version of Paxton's Lily House at Chatsworth – the transept seemed to rise to fairy-tale proportions. According to one eyewitness: 'its vastness was measured by the huge elms, two of which rose far into the air, with all their wealth of foliage as free and unconfined as if there was nothing between them and the open sky'. On entering the Palace, the first sight was of a magical crystal fountain built by Osler, nearly thirty feet high and made of four tons of pure crystal glass. Beyond this, it was possible to glimpse 'the plash of fountains, the luxuriance of tropical foliage, the play of colours from the

The opening ceremony of the Great Exhibition in the
Crystal Palace in Hyde Park, London.

choicest of flowers carried on into the vistas of the nave by the rich dyes
of carpets and fabrics from the costliest looms'.

The eleven miles of stands presented the 'very best that human
ingenuity and cultivated art and science could inspire'. The eastern half
of the Palace was filled with sumptuous displays – silks, tapestries, art,
manufactures – from foreign countries; the western half with products
from Britain and the Empire. Rare and precious *objets d'art*, the latest
ideas in science and technology – the countless exhibits were a symbol
of an optimistic new era when 'commerce and discovery were to bind
the nations of the earth together, and enlightened industry was to
succeed in making an end of war'.

For many months Richard Owen had served on the committee for the
'Great Exhibition of the Works and Industry of all Nations'. This had
brought him into close contact with the royal family; Prince Albert was

personally involved in the Exhibition, and meetings were not infrequently held at the royal palace. A few days before the official opening, Owen was appointed to judge sections of the Exhibition. He was nominated Chairman of 'Jury IV' of the Exhibition, responsible for the division on 'Vegetable and Animal Substances, chiefly used in Manufactures, as Implements or for Ornaments', and a member of 'Jury V' on 'the Animal Kingdom'.

At the grand opening ceremony, 'Richard as juror, took his sister with him in search of the jurors' gallery for which he had a pass', while his family was seated in 'an excellent place, front seats in the central part'. Owen was surrounded by other important dignitaries, fellow-commissioners and jurors such as the MP William Gladstone, the President of the Royal Society the Earl of Rosse, the chemist Dr Lyon Playfair and the Lord Mayor of London William Cubitt. They had an excellent view of the entry of the royal family and the procession that followed.

'The trumpets proclaimed the arrival of the Queen and Prince Albert,' Caroline wrote in her diary. The Archbishop of Canterbury entered first and seemed so astonished at the dazzling array of exhibits that he kept stopping, lost in admiration. As a result, his chaplains could not maintain a stately pace and the lords-in-waiting behind them, who were walking backwards before the royal party, found themselves in danger of stumbling into the clergy. 'Never was a sovereign more heartily welcomed,' Caroline recorded: 'the Queen led the Prince of Wales with her right hand, and her left hand was linked in Prince Albert's arm, who was leading the Princess Royal. Then followed a procession of ladies, and I caught a glimpse of beautiful dresses and diamonds.'

Queen Victoria herself, entering to a triumphant rendition of the 'Hallelujah Chorus', felt awed: 'the glimpse of the transept through the iron gates, the waving palms, flowers, statues and myriads of people filling the galleries and seats around, with the flourish of trumpets as we entered, gave us a sensation which I can never forget and I felt much moved'. It was, she said later, 'the happiest, proudest day of my life'.

Richard Owen, at the heart of the cultivated metropolitan society that

had organised this great triumph, was presiding regally over the arrange
ments. Over the next few days he carried out his duties as juror,
supervising his divisions, entertaining foreign guests and awarding
medals. His very self-assurance seemed to transform anything that was
deemed to come under his care into smooth-running perfection as he
mingled, quite at his ease with senior politicians, ambassadors and
members of the royal family.

It was different for Gideon Mantell. Still struggling with the pain of
his injury and given no preferential treatment, he passed unnoticed in the
crowds, feeling both delight in the scientific presentations and frustration at his own physical limitations. 'The effect is indescribably
overpowering. I cannot express the effect it has left upon my mind,' he
noted in his diary, 'nothing can prepare you for this.' He found himself
irresistibly drawn back many times by the beauty of the science and technology on display at the Exhibition. Every conceivable new invention
and curiosity could be found: the large Ross astronomical telescope,
ornamental clocks – one of which had taken thirty-five years to make –
microscopes, scientific instruments, steam turbines, the latest designs of
barouches and other kinds of carriage; there seemed no limits to the
human imagination. It was, in the words of Alfred, Lord Tennyson, a
'Vision of the World and all the wonder that would be'.

On the 'shilling days', when admission was reduced, people flocked
to London from the provinces by special excursion trains or even on
foot. On one occasion there were over ninety-seven thousand people in
the building, yet, Mantell noted, 'I managed to squeeze into the back and
least crowded compartments of minerals and with some difficulty
ascended the gallery overlooking the transept to look down on the sea of
heads underneath.'

Though frail and worn from the pain he suffered, Mantell would not
stay away, faithfully returning even on the last day of the Exhibition, a
stooped, dark figure in a throng of thousands.

A lovely warm day . . . Remained till the close of the scene and
did not leave the building until nearly half past five [he wrote].

Both to the east and west a sea of black-hats was in constant undulation . . . and so a continuous layer of black was thus formed, the white statuary surmounted the whole in sublime relief like deities . . . When the clock struck five the immense multitude was congregated in the nave, transept, galleries and all was breathless silence. In a moment the organs pealed out the National Anthem . . . Bells, gongs . . . then burst forth and fairly stunned out the lingering multitude . . . the closing scene of the most marvellous display the world ever beheld!

William Buckland took no part in the spectacular proceedings. For some time, he had been showing signs of mental abnormality. To Mrs Buckland's distress, he began to behave irrationally, displaying manic tendencies far removed from his usual considerate and genial character. His strange behaviour intensified almost to the point of violence. He would beat himself about the head and scratch himself roughly 'so as to produce alarm'. The younger children thought their Papa was acting; the older ones were 'horror stricken'. Mrs Buckland was convinced that he was suffering from overwork.

All his life, he had striven to bridge the ever-widening gulf between religion and geology. He had wrestled with opponents of the new science who supported the biblical interpretations, as each new wave of evidence highlighted fresh anomalies. During his career, geologists had shown that the earth was not six thousand years old, but of much greater antiquity. Life was not made in a single week; the six days of Creation had become 'geological ages' covering vast periods of time. There had been no worldwide Deluge; Noah's Flood was increasingly seen as an unimportant regional event and many of the phenomena that Buckland had used to explain it were now thought to be due to glaciation. There was no evidence that creatures had populated the earth from a single site on their release from Noah's Ark. Rather, there appeared to be centres of creation on the different continents. Even the superb design of creatures in which Buckland had seen the hand of God did not fit easily with the progression of life forms in the fossil record. The relentless onslaught of new evidence, the endless attempts to breech an unbridgeable gap, seemed to have finally taken their toll.

As his disturbed and sometimes aggressive outbreaks increased, Mrs Buckland had no idea what to do. She wrote to friends, Owen, Broderip, Murchison and others, for advice. There seemed to be widespread agreement that William Buckland must be removed from his family; the children must not witness such shocking and disturbing behaviour from their Papa. By the time the Great Exhibition was in full swing, Buckland's future was settled. He was committed to a lunatic asylum in Clapham near London, and soon it was said that he was 'lying amongst outrageous madmen'.

Rumours began to fly about the horrific treatments to which he was subject. There was a firm belief that 'the Dean must be brought to reason by the union of control and medical care', and that this must happen before the mental disease takes 'such a root that nothing will remove it'. Letters suggest that Richard Owen supported Mrs Buckland in moving her husband to the mental institution. But others were shocked. 'Oh what a horrible calamity!' wrote Mantell. 'Death would indeed be a relief. It grieves me to see how everything goes on the same in the scientific bodies of which he was so bright an ornament and so energetic a member. No one seems to think of him.'

The collector Thomas Hawkins was so outraged that he wrote to Owen, asking if Buckland was indeed in the Clapham retreat, for 'I should be ready to force the doors of the place open to release a gentleman and scholar like Buckland, while I am filled with indignation to think of such a fate for one who rendered me a good service.' Hawkins, one of Buckland's many admirers, was 'fervently praying to God to assist him in such great affliction'.

It seems likely that Richard Owen acted from the best of intentions in trying to help the family obtain medical treatment for the Dean. Buckland had, after all, been a close ally and friend for more than twenty years, and his departure from the world of science could only be a loss. But even without his old patron, Owen now enjoyed the attentions of such a constellation of eminent supporters that his career was assured and his fame was beginning to reach Europe.

After the Great Exhibition, the President of the French Republic invited him to Paris, where he was nominated as a grand juror for the planned Universal Exhibition. Escorted by Hussars and Dragoons, Owen and the other jurors enjoyed the very best of French hospitality, at Versailles, at the Opéra, at a splendid banquet in the Orangerie of the Jardin des Plantes. On his return, at the request of Prince Albert, he was asked to give a talk at the Royal Society of Arts on his section of the Exhibition. More honours followed when the King of Prussia decorated him as a Chevalier of the Ordre Royal pour le Mérite dans les Sciences et les Arts.

Despite his meteoric rise and unshakeable position, Owen still persisted in his hostilities against Gideon Mantell. In a final, bitter dispute – 'which worried Mantell to death', according to one reporter – they fought over an extinct reptile found in strata known as the Elgin sandstone of Morayshire. The Elgin rocks were thought to be part of the Old Red Sandstone formed in the Devonian period. If this was the case, the tiny lizard would be the most ancient yet uncovered. Mantell was one of the first to hear the news, since the discoverer, Patrick Duff, was a relative of his friend Captain Brickenden. To add to the excitement, the Captain had recently come across turtle footprints embedded in the same stone near Elgin. All this suggested that reptiles had inhabited the globe for much longer than was previously thought.

Word spread fast of the anomalous find. Charles Lyell was reported to be 'inebriate with joy'. The discovery of a reptile in strata in which they had not been uncovered before lent weight to his view that the fossil record was unreliable and haphazard. Lyell, like Owen, was convinced the progressionists were wrong. He envisaged this as the first of many skeletons still buried which would show that any apparent order in the fossil record over time was a myth. Patrick Duff made arrangements for his brother George in London to show Lyell and Mantell the strange new creature.

The shape entombed in the sandstone was barely six inches long. with a curved tail and splayed-out limbs. It seemed amphibian-like, similar to salamanders and newts, yet with reptilian characteristics in the arrangement of the bones in the skull, the roof of the mouth and the vertebrae. Gideon Mantell wrote at once to Captain Brickenden: 'though I had but a transient glimpse of it, yet I saw enough to reveal its general character . . . the reptile is a very primitive one . . . I propose naming it *Telerpeton elginese*, from the Greek signifying the remote or most ancient reptile, a very pretty name is it not?' They planned to present a joint paper to the Geological Society in December 1851, announcing the discovery of reptiles in the Elgin sandstone of Morayshire.

But Professor Owen was also well aware of the exciting breakthrough. He, too, came to the meeting at the Geological Society just before Christmas. Unfortunately for Mantell, other matters on the agenda occupied the whole evening and he had no chance to describe his study, which was deferred until the next meeting. He was frustrated to see Professor Owen carefully examining the enigmatic fossil and his own anatomical drawings, which were on display. A couple of days later, Owen rushed into print in the *Literary Gazette*, 'prigging Mantell's bones', according to a later report. He classified the new creature as a lizard. In characteristic style, he ignored the name Gideon Mantell had assigned and named the creature himself: '*Leptopleuron lacertinum*' or 'slender-ribbed reptile'.

Mantell was horrified. 'It really is very sad, after the labor I have bestowed in working out the story to be subject to such annoyance,' he told Brickenden. To complicate matters, Owen claimed that he had been asked to name and describe the new creature. Soon after this, the *Literary Gazette* misquoted a report of a lecture by Charles Lyell, so that it appeared that Lyell recognised Owen's name for the creature, not Mantell's, when in fact the reverse was true. Lyell protested to the *Gazette*: 'those who heard my lecture well know that I said nothing of Mr Owen's opinion on this matter.' Instead, Lyell pointed out, 'I exhibited a cast which Dr Mantell had made of the reptile . . . on which was inscribed "*Telerpeton Elginese*".'

Gideon Mantell did have an opportunity to read his paper to the Geological Society, during the first week in January. It soon transpired that Professor Owen had written to the President demanding that his account in the *Literary Gazette* should supersede Mantell's, since it came first. Archives show that, for once, there may have been some justification in Owen's claim. Dr George Duff, in his enthusiasm to gain recognition, had shown the specimen to Lyell, Mantell *and* Owen. Owen believed that he had been asked to describe the creature.

Whatever the truth of the matter in this case, it would appear that members of the Geological Society had had enough of Owen's manoeuvres. According to Mantell, 'Owen's conduct created a vehement outcry against him.' He told Brickenden, 'the unanimous decision was, of course, in my favour, and the President announced from the chair most emphatically that our paper was, to all intents and purposes, published at the former meeting and must take precedence of any other'. The President was 'warmly applauded'.

Owen himself was conspicuously absent from this particular meeting at the Geological Society. In the midst of the furore over *Telerpeton*, he received an unexpected and almost unprecedented honour for a scientist. A letter arrived for him at Lincoln's Inn Fields from the royal palace at Osborne, with the Queen's arms embossed in black wax. It was from her assistant, Mr Phipps:

> My dear Sir,
> I have been commanded by the Queen to inform you that, a house upon Kew Green having become vacant by the death of the late King of Hanover, Her Majesty is happy in being able to offer this house as a residence for you . . . The Queen commands me to say that she thinks that there is no method in which she can better give a tribute of her respect and regard for science than by thus meeting what she believes to be the almost necessary convenience of one of its chief ornaments and most distinguished members.

Jubilant, Owen wrote to inform his sister that, after all this 'medal getting' and 'foreign orders of Knighthood', now comes the 'solid pudding'. But as news of Owen's good fortune spread, it provoked further jealousy. At one dinner engagement, Owen wrote, he was 'attacked about his palatial residence' by Sir Robert Inglis, a Tory politician and trustee of the British Museum. Finding that the rights of possession of Her Majesty's Kew residence were in dispute with the family of the King of Hanover, Owen quickly found a large, rambling house in Richmond Park called Sheen Lodge, which belonged to the Queen and was also empty. This, he thought, would suit his purposes admirably and, being slightly smaller, would perhaps attract less criticism.

Owen travelled to the Isle of Wight to see Prince Albert at Osborne. He found the Prince 'planning out the grounds so as best to instruct his children in botany, and he asked Owen's advice as to the best method of so doing'. Owen explained to him his interest in the Queen's house at Richmond. His wishes were conveyed to the Queen, and Richard Owen soon found himself a man of considerable property. On his first night in his grand new residence he slept till late, 'in comfortable, instinctive unconsciousness that the whole was a reality and no early morning dream'.

Enveloped in the English countryside and now elevated well beyond the harm of petty criticism, Owen still could not resist airing his grievances and tormenting his rival. 'In the last *Literary Gazette*,' Mantell noted on 27 January, 'there is another attempt to establish Professor Owen's priority of description and name of *Telerpeton*.' He was also concerned at Owen's manoeuvres at the British Museum which, if successful, would give him direct control over Mantell's collection. A senior post at the museum became vacant when Charles Konig, the Keeper of Mineralogy and Geology, died suddenly. Owen quarrelled bitterly with Lyell, who refused to recommend him for the post. As it gradually emerged that Owen did not have enough support, he was forced to withdraw his application. 'At all events my collection is safe from his clutches,' Mantell wrote with relief.

In the summer of 1852, Mantell finally received the opportunity that meant so much to him. The Crystal Palace Company aimed to relocate the Great Exhibition on a permanent site in two hundred acres of landscaped grounds on Penge Hill, Sydenham, just south of London. A few acres of parkland were to be devoted to geology. The board of directors approached Mantell to see if he would oversee an ambitious project: the first life-sized restorations of the dinosaurs.

At a meeting in August, the directors of the company had decided that a 'Geological Court be constructed containing a collection of full-sized models of the Animals and plants of certain geological periods, and that Dr Mantell be requested to superintend the formation of that collection'. The minutes of the meeting show that they wished to make 'enquiry of Dr Mantell what degree of completeness such a collection could attain for a sum of 3000 or 4000 pounds'. Here, at last, was the recognition for which Mantell had longed. This was his chance to conjure into being his vision of the extraordinary monsters, *Iguanodon* and the other dinosaurs, to which he had dedicated his life.

But the commitment required at least a year's work. Now emaciated and almost 'frantic with pain', he had become so used to taking opiates that he could swallow an ounce of liquor opii sedativus at a time, thirty-two times the maximum dose! His nights passed in agony, with 'neuralgia flying from one limb to another . . . no relief from prussic acid, liniments, fomentations, calomel, and opium, hot brandy etc all no avail'. Mantell knew he was dying, and he declined the honour. 'Very good for nothing,' he wrote. 'In truth, I am used up.' Hearing of his plight, Earl Rosse, the President of the Royal Society, made arrangements for Queen Victoria to give him a civil list gratuity of £100 a year in recognition of his scientific labours.

A few months later, on 10 November 1852, Mantell slipped on the stairs at home and 'was obliged to crawl upon his hands until he reached his bedroom'. He took half a dose of opiates, but when this had no effect, he finished the dose two hours later. The next day he died of narcotic poisoning. He was buried in Norwood, beside his 'beloved child' Hannah Matilda. In accordance with his wishes, the funeral was as 'plain as possible' and no one was invited to attend.

A post-mortem revealed the extent of Mantell's spinal deformity. The lower part of his backbone had 'a remarkably twisted appearance'. Five lower vertebrae were affected — some had rotated so far that they were almost at right-angles to their correct position. Owing to this extreme curvature, the bony transverse processes of the vertebrae were now twisted and projected both outwards and into his abdomen. It was these protrusions that had been misdiagnosed as a tumour or abscess. The discs and cartilage separating the vertebrae had been virtually destroyed.

Ironically, and perhaps a little ghoulishly, the specimen of Mantell's damaged lower spine was sent to the Royal College of Surgeons, where it was placed in Owen's museum. His broken back was to become a pathological specimen in a bottle of preserving fluid on a shelf in the Hunterian Museum, a unique exhibit to illustrate 'the severest degree of deformity of the spine'. But not even this final victory, with his rival's remains dissected, preserved and classified in a suitably scientific manner and now totally under his control, put an end to Owen's opposition.

With Gideon Mantell scarcely cold in his grave, an anonymous obituary was published in the *Literary Gazette*. In it, Mantell was dismissed as an inadequate scientist, 'in want of exact knowledge'. The unidentified writer talked at some length of his 'weaknesses' and his 'overweening estimate' of his own importance, which the writer 'had occasion to deplore'. Even the discovery of *Iguanodon* was taken from him. According to the obituary: 'To Cuvier we owe the first recognition of its reptilian character, to Clift the first perception of the resemblance of its teeth to those of the Iguana, to Conybeare its name, and to Owen its true affinities among reptiles, and the correction of errors respecting its bulk and alleged horn!'

Leaders of the geological community were shocked, and in no doubt that the author was Richard Owen. 'Have you seen the article in the *Literary Gazette*?' William Hopkins, the current President of the Geological Society, wrote to his friend Leonard Horner; 'I think it is palpably from Lincoln's Inn Fields. It bespeaks a lamentable coldness of the heart of the writer.' Owen was denied the presidency of the Society that year, in view of his 'pointed and repeated antagonism to Gideon Mantell'. Hopkins had written to a colleague explaining that he felt very divided about Owen's eligibility, since 'I should feel it scarcely respectful to the memory of poor Mantell to nominate one whose occupancy of the Chair would have driven him so entirely from the Society had he been living.' Owen, in fact, never became President of the Geological Society.

To many of his colleagues, Owen's outburst against Mantell seemed badly misjudged. Owen's own reputation was secure. He was, rightly, acknowledged as Britain's foremost anatomist and an international authority in his field. Yet unlike Mantell, who had been compelled to earn his living as a country doctor, Owen had been fortunate enough to be able to devote his working life to the subject he loved. By the time of Mantell's death, Owen's breadth and depth of knowledge of anatomy far surpassed Mantell's. Yet Owen's achievements and international acclaim seemed to unleash an even greater, almost fanatical, egoism and a callous delight in savaging his critics. Although Mantell's legacy posed no threat to Owen's eminence, his death provided an opportunity for him to display a sadistic streak that was needlessly channelled into crushing Mantell's reputation.

Many others, however, were inclined to be more generous. William Hopkins, in an address to the Geological Society described Gideon Mantell as 'a memorable instance of a man of genius', who 'attained great eminence as a man of science' despite being 'constantly and diligently occupied with the practice of a laborious profession'. The *Illustrated London News* highlighted his skill as a lecturer: 'Dr Mantell took great delight in imparting to others a knowledge of his favourite science. He was fluent and eloquent in speech and full of poetry.' According to his colleagues at the Clapham Atheneum: 'no one who has enjoyed the advantage of hearing him . . . can ever forget the singular ability . . . and the energetic eloquence which characterised his discourses'. *The Gentleman's Magazine* wrote of his brilliance as a discoverer and a collector. For Benjamin Silliman the loss was of an outstanding personal friend whose candour, kindness and intellectual ability he had appreciated for many years: 'Exact and thorough scientific knowledge, the enthusiasm of a discoverer and the rich but chastened diction of a poet were never more remarkably united than in him.'

Reginald Mantell, who had been working in America, resigned his post and returned as soon as he heard the news of his father. He wrote to his cousin, 'I had not taken the key out of the envelope, feeling as I did a sad dislike to re-enter those rooms which I remembered as warmed by a father's love, and which are now, alas, so cold and desolate.'

Mantell's few remaining possessions were dispersed. Five hundred pounds to Walter; books, fossils and antiquities to Reginald; £50 to the maid-servant who had helped to nurse Hannah; a pair of ancient china jars to a friend. Reginald, under the terms of the will, sold some of Mantell's remaining fossils to the British Museum – a hundred verte-brates and over a thousand invertebrates. It took several months to carry out his father's wishes; then, the following year Reginald set sail for India to find work as an engineer in the Empire.

Only after Mantell's death did Owen finally accept his errors in in-terpreting the belemnites, but he was never explicit. It was quite beyond him to withdraw his criticism of Mantell, or publicly admit that he had made a mistake. Yet he began to refer to *Belemnoteuthis* as the 'apparently guardless species', indirectly acknowledging the correctness of Mantell's view. There were other revisions too. It wasn't long before the name *Telerpeton elginese*, which Mantell had so proudly given to the primitive reptile from Morayshire, disappeared with a reclassification by Owen. The Elgin reptile became known by Owen's name, *Leptopleuron*, which he insisted had priority.

Almost inevitably, the opportunity to reconstruct the dinosaurs for the permanent exhibition at the Crystal Palace at Sydenham fell into Owen's lap. He collaborated with Benjamin Waterhouse Hawkins, Director of the fossil department at the Crystal Palace, and they began by designing miniature models in clay, according to Owen's vision. When Owen had checked these for accuracy, Hawkins and his team created life-size clay models, some weighing more than thirty tons. A mould was then made of each dinosaur, and used to form a huge metal cast. Gradually, in Hawkins's workshop at the Crystal Palace, an array of monsters came slowly into being: giant sea lizards, *Ichthyosaurus*, *Plesiosaurus*, pterodactyls (the flying lizards) and the dinosaurs, *Megalosaurus*, *Iguanodon* and *Hylaeosaurus*.

Dinosaurs under construction at the Crystal Palace in Sydenham,
from the *Illustrated London News*, 1853.

The model of the *Iguanodon* itself was a stupendous feat. Some thirty-five feet long, it was constructed with 'four iron columns, nine foot long, seven inches in diameter; 600 bricks, 650 five-inch drain-tiles, 900 plain tiles, 38 casks of cement, 90 casks of broken stone . . . These with 100 feet of iron hooping and 20 feet of cubic inch iron ore constitute the bones, sinews and muscles of this large animal.' Ignoring Mantell's correct suggestion that the animal's forelimbs were smaller and used for seizing and grasping, Owen reconstructed his own vision of *Iguanodon*: a four-footed brute with stumpy, pillar-like legs, a squat, bulky body and thick-scaled skin. Never before had such an ambitious reconstruction of an animal been attempted.

To promote their efforts, while they were making the *Iguanodon*, twenty-one distinguished guests were invited to a banquet inside the belly of the beast on New Year's Eve 1853. The intriguing invitations were written on the outstretched wing of a drawing of a pterodactyl: 'Mr Waterhouse Hawkins requests the honor of —— at dinner in the mould of the Iguanodon at the Crystal Palace on Saturday evening December the 31st at five o'clock 1853 – an answer will oblige.'

As a publicity stunt it was superb, a triumph of showmanship. Surrounded by the half-formed skeletons of the gigantic creatures, which were supported by great hoists, ropes and pulleys, eleven dignitaries, in the full splendour of Victorian evening dress – white cravats, jewelled pins, gold chains – were seated at a table punctiliously set for an eight-course meal with accompanying wines, inside the belly of the *Iguanodon*. Ten more sat at a table placed alongside. The snowy white linen, the dark suited shapes of the guests, the curious shadows cast by the suspended lamps, the claustrophobic feel engendered by being inside the restricted space of the creature's stomach – it was all made the more bizarre by the great swathes of pink and white brocade used as an awning around the prehistoric animal. Draughtsmen from the *Illustrated London News* hovered, eager to record the unique sight of the leaders of science feasting in such a novel setting.

This was the giant creature that Gideon Mantell had struggled to define for almost thirty years. The gentlemen were surrounded by models of fossils that he had been the first to chisel out of the rock, night after night, sacrificing his marriage, his health and his professional practice to the uncovering of the ancient past. Yet it was Richard Owen, the man who had opposed and thwarted him so often, who sat as guest of honour in the most eye-catching position in the head of the beast, receiving the credit for interpreting *Iguanodon*. The restorations were seen as a masterpiece: 'the highest point of knowledge [of the great dinosaurs] which had been attained up to the present'. Praise was heaped upon the 'Newton of Natural History', who presided royally over the event, leading the speeches and ostentatiously celebrating his victory over Mantell.

The only record of Mantell's labours was a small plaque attached to the awning, in the shadows above the tail-end of the beast, placed, perhaps significantly, directly opposite Richard Owen. There were plaques, too, for Cuvier and Buckland. Yet in his speech, Owen, glorying in the occasion, took the opportunity to applaud his own team's work. 'It has been a great source of pleasure to aid so important an undertaking,' he said 'by assisting with instruction and direction a gentleman who possesses the rarely united capabilities of an anatomist, naturalist and practical artist . . . which has ensured Mr Hawkins's careful restorations the highest point of knowledge which has been attained up to the present period.'

Then the chairman of the Crystal Palace Society rose to congratulate Richard Owen on 'the great interest evinced and approbation expressed by H.M. the Queen and H.R.H the Prince on their recent visit to the extraordinary works by which the company were surrounded'. Comparing Owen to Cuvier, he continued, 'the restoration from a single fossil fragment to complete skeletons of creatures long since extinct, first effected by the genius of Cuvier, has always been considered one of the most striking achievements of modern science. Our British *Cuvier*, Professor Owen has lent us his assistance in carrying these scientific triumphs a step further and in bringing them down to popular apprehension.'

'After several appropriate toasts . . . this agreeable party of philoso-

'A Dinner Party in the *Iguanodon*' from the
Illustrated London News, 1854.

phers . . . were evidently well pleased with the modern hospitality of the
Iguanodon,' declared the local paper with some understatement. In fact,
after sherry, Madeira, port, moselle and claret, the learned gentlemen
were moved to sing, with everyone joining in a rowdy chorus:

> 'The jolly old beast
> Is not deceased
> There's life in him again! [a roar]'

The party did not break up until well after midnight. According to
Hawkins, as the drunken group made their way across the park to the
railway station, 'the roaring chorus' was 'so fierce and enthusiastic as
almost to lead to the belief that the herd of Iguanodons were bellowing'.
Details of the theatrical scene and of Owen's fantastic achievement
were soon portrayed even in papers in Europe. The *Illustrated London*

News, hitting a suitably restrained note, applauded the novelty of the banquet inside the *Iguanodon*'s 'socially loaded stomach'. The occasion, they said, had 'excited the curiosity of the leading scientific men of the country', and they were 'evidently well pleased with the modern hospitality of the Iguanodon'. The *Punch* reporter, under the headline 'Fun in a Fossil', pointed out that Professor Owen and his friends had 'an exceedingly good dinner . . . Had it perhaps been an earlier geological period they might have occupied the Iguanodon's inside without having any dinner there.' The *London Quarterly Review* was inspired: 'Saurians, Pterodactyls all! . . . Dreamed ye ever . . . of a race to come dwelling above your tombs and dining on your ghosts.'

Neither William Buckland nor Gideon Mantell, of course, were to witness the captivating impact of their extraordinary discoveries on the world of the 1850s. Richard Owen was courted by popular magazines and urged to write for them. Charles Dickens pleaded with him to begin a series of zoological articles for his journal *Household Words*. 'It would be in vain for me to attempt to tell you with what pride and pleasure I should receive such assistance,' Dickens wrote, 'or what high store I should set by it.' Dinosaurs made a fleeting appearance in his novel, *Bleak House*, of 1852: 'Implacable November weather. As much mud in the streets as if the waters had but newly retired from the face of the earth, and would it not be wonderful to meet a Megalosaurus, forty feet long or so waddling like an elephantine lizard up Holborn Hill.'

Even before Waterhouse Hawkins's dinosaurs were finished, they were a sensation. He was 'besieged' with requests to view the monsters taking shape in his studio in the grounds of the Crystal Palace. The Palace itself was being rebuilt on an even larger scale than the original, and was surrounded by landscaped gardens with a 'display of fountains more than four times that of Versailles'. Yet according to the *Times*, 'if more was wanted to astonish' it could be found in 'the gigantic tenants of this planet before Man's introduction upon the scene'.

Such was the interest that by 10 June 1854, when Queen Victoria opened the permanent Exhibition, forty thousand spectators flocked to the Crystal Palace at Sydenham. Richard Owen arrived in the company of the Prince Consort, the French Emperor and the King of Portugal. Two huge *Iguanodons*, a *Megalosaurus* with spiky teeth and a *Hylaeosaurus* with dagger-like spines were displayed on an island, showing the Mesozoic era. There were pterodactyls on a crag above, and plesiosaurs and ichthyosaurs rising up out of the water. These, the world's first prehistoric sculptures, were 'irresistible to the public'.

The popular understanding of dinosaurs was, inevitably, still very limited. Even though the idea of a Flood had been discredited in geological circles, Owen's creations became integrated into the biblical history of the earth and were widely accepted as beasts that were destroyed in Noah's Flood. In the words of the *Westminster Review*, some even believed these 'savages and beasts' had perished 'because they were too large to go into the Ark'. Following Owen's endeavours, *Iguanodon* and *Megalosaurus* were usually portrayed as clumsy, rhinocerine, four-footed reptiles with very heavy proportions. From the fragmentary evidence then available, no one could envisage the large number of different dinosaurs that remained to be found. Above all, they were not thought of as the product of evolution, but as 'created' in some way by a wise God who saw them as the most fitting creatures for the earth in its infancy.

In the next decade, hundreds of thousands of visitors went to view Owen's 'Mausoleum to the memory of a ruined world'. Models and posters of the Crystal Palace dinosaurs were widely distributed, and inspired popular literature and drawings. Jules Verne in *Journey to the Centre of the Earth* depicted an *Ichthyosaurus* and *Plesiosaurus* locked in mortal combat, and Louis Figuier's *The World before the Deluge* of 1863 gave a vivid portrait of dinosaurs fighting in the antediluvian era. Dinosaurs became a regular feature of *Punch* magazine. Benjamin Waterhouse Hawkins became the foremost natural history illustrator, and was requested to create a lifesize prehistoric museum in Central Park, New York. As the first wave of 'dinomania' took off, the mantle of glory for the single-handed discovery of dinosaurs fell on Richard Owen's shoulders, and the public and the establishment adored him.

The efforts and sacrifices of those intellectual giants on whose shoulders Owen stood were gradually forgotten. William Buckland, for so long the fair-minded mediator in disputes with colleagues, eager to raise subscriptions for Mary Anning or to write letters on Gideon Mantell's behalf, could no longer raise his voice to ensure credit was fairly given. None of the medical treatments had stemmed the onslaught of his inexplicable disease. Yet he lingered on for several years, the terrifying degeneration of that once clear mind slowly destroying all that was recognisable of him.

There was no way of knowing what occupied his thoughts. When favourite objects of natural history were placed in his room he would show no interest, or violently dismiss them. Close friends or relatives, too, produced little response: 'The Dean would not speak to my uncle and looked another way,' observed his eldest son, Frank. 'He would answer no questions and make no remark and seemed glad when my uncle took his leave.' Nothing it seemed could rouse his interest, except the Bible. Mrs Buckland resolutely retained her optimism that he would recover: 'he is afraid of trying his legs, I think, so he rests on Frank's neck and manages pretty well to ascend the steps that lead into the room'. But it wasn't long before even this was beyond him, and he died in 1856.

Soon after this, Mantell's younger son, Reginald, also died. He had tackled major engineering projects in India and survived a mutiny, only to fall victim to cholera. He was just thirty when he was buried in Allahabad. Upon hearing the news, Walter Mantell made arrangements to return to England from New Zealand and gathered together his father's remaining fossils and antiquities. He visited Sir Charles Lyell, who helped him to label the specimens. They even identified the first *Iguanodon* tooth which Lyell, as a young man, had presented hopefully to Cuvier. Two cases of fossils from Sussex, along with other prized possessions, were transported back to New Zealand. Charged now with sole responsibility for his father's heritage, Walter Mantell helped to found a scientific society in Wellington, which became the Royal Society of New Zealand.

In England, all that remained of Gideon Mantell's Herculean efforts was his collection in the British Museum. But by the spring of 1856, arrangements were being made to create a special post for Owen at the Museum, as 'Superintendent of the Natural History Department'. As Lord Macaulay wrote to his fellow-trustee there, the Marquis of Landsdowne, 'Owen's fame was spread over Europe . . . He is an honour to our country . . . and a case for public patronage.' Others agreed with this view: 'we have a magnificent collection in the British Museum and an unrivalled expositor in Professor Owen, why are the two separated?' The branches of geology, zoology, botany and mineralogy were to be upgraded to individual departments, all of which would fall under Owen's jurisdiction. The princely sum of £800 a year was offered with the post, but Owen scarcely needed such enticement.

The fossils buried for millions of years, painstakingly scoured from the earth by Gideon Mantell and many others, were now finally delivered to Owen's care. But it was obvious to him that the crowded, damp conditions in Bloomsbury were inadequate to display the wonders of the natural world. He did not hesitate to promote his cherished scheme for a national museum of natural history. A brilliant lobbyist, he now had the connections to ensure his voice was heard. After discussions with the Prince Consort and William Gladstone, a rising star in the Liberal Party and soon to be Chancellor of the Exchequer, Owen formally submitted his plans to the Treasury in 1859. His elegant drawings set out his wishes for the three kingdoms of Nature: plants, animals and minerals. He hoped for nothing less than a ten-acre site in the heart of London in which almost every species of higher animal would be displayed. With great enthusiasm, Owen even proposed a ninety-foot whale gallery.

It was to be 'the best and noblest museum in the world', for lovers of natural history. 'Every organism is a character in which the Divine wisdom is written,' he wrote. As he once told his son when he had found a tiny sea creature on the beach, 'both we and it are the works of a great Creator who never loses sight of the working of his machines. Let nothing disturb your feelings of reverence for Him when in His house and engaged in his worship.' For Richard Owen the ambitious scheme was nothing less than a monument to God.

14

Nature without God?

Things fall apart; the centre cannot hold;
Mere anarchy is loosed upon the world,
The blood-dimmed tide is loosed, and everywhere
The ceremony of innocence is drowned . . .

William Butler Yeats, *The Second Coming*

No sooner had Owen placed his proposal before the parliamentary committee than he found himself, arguably for the first time in his life, outflanked by younger colleagues. Growing concerns over his immense power and patronage erupted into a hostile campaign against his cherished scheme for a natural history museum.

One of his most vocal opponents was Thomas Henry Huxley, who after a four-year struggle to obtain a scientific post had finally become a lecturer in 1854 at the Government School of Mines. Brilliant and combative, Huxley had used his platform to launch himself on to the scientific stage. As a naturalist for the government's Geological Survey he became an expert on vertebrate fossils, and was soon appointed to the prestigious position of Fullerian Professor at the Royal Institution. Professor Huxley viewed Owen's plans for a natural history museum as an extension of his insidious control of science, the 'temple' from which Owen as the 'Autocrat of Zoology and Palaeontology' sought to dominate the field.

Thomas Huxley was so determined to curb Owen's power that he lobbied the Chancellor of the Exchequer, pointing out that a new museum was unnecessary and would cost a 'prodigious sum of money'.

Far from ten acres, two would be quite sufficient. As for Owen's absurd extravagances such as a ninety-foot whale gallery, this would make an 'intolerable stench'. Owen's scheme, he told a select committee in 1858, was 'little matured', and certainly 'would not be convenient, either for the man of science or the general public'. An alternative plan was proposed by Huxley and his allies, in which the natural history collections could be dispersed. Fossil plants could be placed in Kew Gardens, where Huxley's friend the botanist Joseph Hooker worked as Director, and minerals in the Museum of Practical Geology, where he himself was Curator.

But as the parliamentary committee rumbled on, considering the evidence, an even greater challenge, thirty years in the making, lay in wait for Owen. Charles Darwin's *Origin of Species* rolled off the presses in 1859, with a modest print run. Even the publisher did not anticipate the furore that would ensue and the shock waves that would ricochet through the gentrified world of science, sweeping aside long held values, and all who cherished them, including the apparently invulnerable Richard Owen.

Darwin had watched from the sidelines for years, as scholars had wrestled with the growing body of conflicting evidence between geology and the Bible. He had been profoundly influenced by Lyell's book *The Principles of Geology* when voyaging as a naturalist between 1831 and 1836. Compared to Cuvier's inexplicable catastrophes, Lyell's concept of slow and steady geological change seemed to fit everything he had observed. 'The great merit of *Principles*,' Darwin had said, 'was that it altered the whole tone of one's mind.' He had applied Lyell's rigorous approach to his own observations. The prevailing view was that species, once created, did not change over time. After all, the progressionists – for all their efforts – had failed to show one species changing into another in the fossil record. But as Darwin had studied an immense variety of species on his travels, he was puzzled by certain anomalies.

Why did species on oceanic islands resemble those of neighbouring continents, with African-like species in the Cape Verde Islands and South American-like species in in the Galapagos Islands? Since the Galapagos and Cape Verde Islands had similar physical conditions, why didn't God create the same animals for them both? Odder still, why would the Creator design different plants for the different sides of a mountain? The vegetation on the east and west sides of the Andes, for example, differed, although the soil and climate were similar.

The Galapagos Islands opened his eyes. Each island had species of animals peculiar to itself and yet related to creatures on the other islands. There were thirteen different species of finch alone, each one with a different size or shape of beak. Why would God make different finches for each island? Such localism seemed absurd. How much more logical to assume that in the slightly different environments of each island, species had evolved along separate lines from a common ancestor.

But, if so, what was the mechanism? How would new species arise in these different environments? On his return to England, Darwin had turned to conventional breeding to consider how new varieties such as the tumbler pigeon or the racehorse were formed. Domestic animals could be bred to favour certain characteristics of size or shape by breeding only from offspring that had the desired traits. By repeating this 'artificial selection' over many generations, the pigeon-fancier or horse-breeder created new breeds. Could the same process occur in nature? How did Nature select?

In 1838, Darwin had read Thomas Malthus's *Essay on the Principle of Population*, which showed how population growth is constantly held in check by limited resources. While man's population had the potential to double in twenty-five years, food supply could not increase so fast, resulting in famine and death. Darwin had applied this reasoning to Nature. Animals, too, were in competition over limited resources. 'Owing to this struggle for life, any variation, however slight,' Darwin wrote, 'if it be in any degree profitable to an individual of any species . . . will tend to the preservation of that individual and will generally be inherited by its offspring.'

In each ecological niche, competition to survive between organisms would favour any characteristic that could give the offspring an advantage. Darwin reasoned that over time, these advantageous characteristics would increase, resulting in an animal that was very different from its remote ancestors. 'I have called this principle, by which each slight variation, if useful, is preserved, by the term Natural Selection,' he wrote, 'in order to mark its relation to man's power of selection.' Just like the 'artificial selection' of the breeder, 'natural selection' could, over countless generations, result in new forms. 'Natural Selection is a power as immeasurably superior to man's feeble efforts,' Darwin observed, 'as the works of Nature are to those of Art.' Far from a Creator miraculously creating all the different species, the infinite variety of life could be explained by the principle of 'Natural Selection'.

Charles Darwin was well aware of the debates that permeated the Geological Society and the Royal Society in the 1830s and 40s; the turmoil created as each new piece of evidence challenged ideas on a biblical Flood, the age of the earth, the order and time-scale of Creation. He watched as even those he admired, like Charles Lyell, tried to explain away the apparent progression of life in the fossil record. All the while, Darwin not only accepted it, he also had a principle to explain what could be driving evolution.

As vividly shown by his biographers Adrian Desmond and James Moore, Charles Darwin was so 'tormented' by the implications of his ideas that he retreated to the rural life of a semi-invalid, 'writhing on his sick bed, fearing persecution'. His views implied that Man was no longer specially created by God, but might have evolved from apes. Few scientific leaders seriously believed that Man could be the product of evolution, except Darwin. If he was correct, intelligence and morality were little more than mere accidents of nature; he 'trembled at his innermost thoughts', viewing himself as 'the Devil's Chaplain'.

For years he refined and extended his arguments in secret. He could not bring himself to confide in friends such as Lyell until 1856. Then, despite Lyell's encouragement, he hesitated to make his views public until he received a letter in 1858 from another scientist, Alfred Russell Wallace, setting out evidence for evolution. Even then Darwin was reluctant to publish. He read his proofs for Origin of Species 'amid fits of vomiting'. In the run-up to publication in 1859, he described himself as

Charles Darwin, 1854.

'living in Hell'. Fearing the attention that might follow, he took like-minded scientists into his confidence: Joseph Hooker, the Director at Kew Gardens, and Thomas Henry Huxley at the School of Mines.

Huxley was 'surprised' by 'the greatness of the book' as he read his pre-publication copy. Anticipating the furore, he wrote to reassure his worried friend: 'I am sharpening up my claws and beak in readiness.'

In the immediate aftermath of the November publication, Charles Darwin was extremely anxious to know what position Richard Owen would take on his ideas. Even friends of the family wrote to enquire about Owen's verdict. 'Dead against us, I fear,' Darwin replied. To his relief, Owen's immediate reaction was not hostile, but ambiguous to the point of even seeming favourable.

When he met Darwin in December 1859, Owen praised him for his original ideas on the formation of species. Owen did not accept that Man was a transmuted ape, but in the *Origin* Darwin had only hinted at Man's relationship with apes. He was eager to build bridges with the famous anatomist, and it is possible that Owen may have imagined that there was some common ground between them: each step in Darwin's evolution could still be planned by God. Behind the scenes, Owen even capitalised on the excitement generated by *Origins* to promote his own cherished aims for a natural history museum. 'The whole intellectual world this year has been excited by a book on the origin of species,' Owen reported to a parliamentary committee. 'Visitors come to the British Museum and they say, "Let us see all these varieties of pigeons: where is the tumbler, where is the pouter?" and I am obliged to say with shame, "I can show you none of them."'

But in the ensuing months, Owen, who had just been knighted, received a series of objections from religious and scientific leaders. The Reverend Adam Sedgewick in Cambridge was one of many who were outraged, even anguished, at the materialistic implications of natural selection. It was inconceivable to him that new species, including Man, arose from a series of random events in nature, not God's will. According to the Church, God had created Man in his own image. If Darwinism were a correct theory, this would imply that God was an ape – an utterly blasphemous idea. The explorer Livingstone, the Duke of Argyll, then the Lord Privy Seal and other leading politicians such as Gladstone, looked to Owen for guidance on the new biology. Even Charles Lyell, who abhorred 'parson-led' science, was opposed to anything 'which tended to break down the barrier between Man and the rest of the animal world'.

While Darwin retreated to the sanctuary of his home at Down House in Kent, Thomas Huxley shrank from no opportunity to spell out clearly the implications of *Origins*. In his Royal Institution lecture of February 1860, before a distinguished audience including Owen, he discussed Man's relationship to the apes, highlighting similarities. Owen was furious; he had always sought to show that Man was zoologically distinct from the animals. Those seeking to reconcile the findings of science with their belief in the Bible faced a terrible dilemma. How could the 'monkey theory' fit with Creation in Genesis? Owen was the obvious scientific leader who could surely be relied upon to expose the flaws in Darwin's thinking: 'The high authority of Professor Owen in the scientific world renders every deliberate opinion pronounced by him a matter of importance,' declared the *Manchester Spectator*.

As the furore continued, Owen could no longer shelter behind the ambiguous language he had used for so long. The *Manchester Spectator*, in reviewing one of his lectures as early as 1849, had highlighted the fact that he appeared to believe in natural laws: 'Richard Owen undertakes to demonstrate scientifically that the arms and legs of the human race are the later and higher developments of the ruder wings and fins of the vertebrated animals . . . he concludes that God has not peopled the globe by successive creations, but by the operation of general laws.'

Following this, Owen had sometimes emphasised 'creative acts' and at other times referred to ill-defined 'secondary' laws of nature, which he thought were preordained by God. As recently as 1858, in his presidential address to the BAAS he had stated that reptiles and mammals were formed on the basis of 'the continuous operation of Creative power', or *'the continuous operation of the ordained becoming of "living things"'* (his italics). But by his own admission he was silent as to the nature or mode of that 'continuous creative operation'. What did this mean? Everyone was waiting for the 'Newton of Natural History' to declare his hand.

It seems likely that, as Owen grappled with Darwin's *Origins*, jealousy may have clouded his judgement. Not known for his generous treatment of rivals, he could only feel eclipsed by the astonishing breadth and clarity of Darwin's arguments. Although his long-awaited review of April 1860 did not explicitly oppose an origin of species by natural law, to the Darwinians it appeared every bit as damning as they had feared.

The original observations, or 'gems', in Darwin's thesis, Owen wrote, were 'few indeed and far apart'. Natural selection failed to explain the 'mystery of mysteries' – the origin of species – any better than existing theories; indeed for the most part it 'rests on a purely conjectural basis'. Where was the proof that all 'the beings that ever lived on this earth have descended, by way of "Natural Selection" from a . . . miraculously created primordial form?' he demanded. Even if there were such proof, this still did not explain the original Creation. Darwin was as mystical in explaining how the first living beings had sprung to life as any of his predecessors.

Owen was even more enraged that Darwin and his 'short-sighted' followers misrepresented alternative 'Creationist' views, feeling that his own ideas of the continuous operation of a secondary law preordained by God had been trivialised. His greatest disgust was for those of Darwin's followers who gleefully espoused the notion that Man might be descended from apes. To anyone who 'deems himself devoid of soul, as the brute that perisheth,' he wrote, such a notion 'may be sufficient and he need concern himself no further about his own relations to a Creator'. But for Owen such ideas were an 'abuse of science', and 'a degradation' of thinking comparable to Lamarck's ideas on transmutation in Revolutionary France. Natural selection, he concluded, had 'frail foundations' and led to 'false philosophy'.

Darwin was worried. He considered Owen's review highly damaging. As a close friend of Prince Albert, and embraced by the powerful Anglican hierarchy, Owen was a powerful enemy. 'It is painful to be hated in the intense degree with which Owen hates me,' Darwin wrote to a friend; 'the Londoners say he is mad with envy because my book has been talked about.' Even though Owen was not a Creationist the sides became polarised, with Darwin and his supporters, 'the Devil's Disciples' Huxley and Hooker, standing in opposition to Owen, who was trying to uphold traditional values.

Their ideological clash came to a head on Saturday 30 June 1860. It took place in Oxford, the home of the clergy and the chosen site for the annual meeting for the British Association for the Advancement of Science. Only twenty years previously Richard Owen had been the undisputed star of the organisation, the chosen protégé of the BAAS. Now, according to the legend in part created by the Darwinian camp, the BAAS meeting was to prove a decisive turning-point for the supporters of the old order.

The Bishop of Oxford, Samuel Wilberforce – uncharitably nicknamed

'Soapy Sam' – was due to talk on botany and zoology. Professor John Draper, of New York University, had been invited to lecture on Darwinism. Richard Owen, who had stayed with the Bishop the night before, was widely believed to have crammed him 'up to the throat' with the best arguments against Darwin. Rumours were flying that the Bishop intended to 'smash Darwin'. This was to be an 'open clash between Science and the Church'. Almost a thousand people crowded into the library to witness the fight. Darwin himself was too sick to attend.

After Professor Draper's contribution, the clergy, 'shouted lustily for the Bishop'. Samuel Wilberforce rose and delivered his address in which, according to Darwin's supporters, 'he said not a syllable but what was in the [Owen's] Review . . . was coached up by Owen and knew nothing . . . ridiculed Darwin badly and Huxley savagely'.

Of the numerous accounts of what happened next, the essence of the story can be distilled. When 'Soapy Sam' had quite finished, he turned grandly to Thomas Huxley and 'begged to know, was it through his grandfather or his grandmother that he claimed his descent from a monkey?' Whereupon, Huxley, it is alleged, 'emphatically striking his hand upon his knee, exclaimed "the Lord hath delivered him into mine hands"'. Slowly and deliberately he rose, and proceeded to outline all the evidence that supported Darwin's ideas. 'The battle waxed hot . . . and the excitement increased,' according to eyewitnesses. Finally Huxley, who was 'white with anger', built up to his shocking finale: 'he would not be ashamed to have a monkey for his ancestor; but he would be ashamed to be connected with a man who used great gifts to obscure the truth'.

According to *Macmillan's Magazine*, 'the effect was tremendous. Lady Brewster fainted and had to be carried out.' As the furore continued, Admiral Fitzroy, who had captained the *Beagle* thirty years earlier, stood solemnly among the audience, raised an immense Bible above his head and 'implored the audience to believe God rather than Man'.

Such dramatic scenes, relayed across England, highlight the outrage felt by leaders of society. Natural science and religion, for so long strained partners, had finally become irreconcilable. Richard Owen, who had earlier embraced the conciliatory path of William Buckland in the *Bridgewater Treatises*, seemed to be trapped, struggling to shore up the values of early Victorian England against the onslaught of evolutionist thinking. All his life, his theories had revolved around a Divine Creator. Even though his views had changed over time and were increasingly ambiguous, he could not contemplate a Nature without the guiding hand of God. For him the works of Nature were, in the words of the psalm, 'telling the glory of God'. Although he did not believe that the Divinity separately made each creature, for Owen the different forms of creation expressed Divine laws.

Now the ground shifted fast, with Darwin's supporters adept at moving on to points where Owen's thinking was most mystical, such as his secondary creative laws. The most forceful and persistent opposition came from Thomas Huxley. Owen had finally met his match. Huxley cornered him on his weaknesses, goading him at every opportunity, the onslaught culminating in a bitter exchange on Man's relationship with apes, which was fought out very publicly, once again in the forum of the BAAS.

Owen had always sought to highlight differences in human and ape anatomy. He aimed to separate Man from Nature by showing the uniqueness of the human brain, 'for the service of the soul'. His argument rested on his claim that there were three anatomical features that were only found in humans: a third, or posterior, lobe in the cerebral hemispheres; structures known as the 'posterior cornu', in the lateral ventricle; and a small internal ridge known as the 'hippocampus minor'. Through his almost exclusive access to the specimens at the Zoological Society Owen had become one of the very few authorities on primates, and for years no one had been able to challenge him.

Huxley began by surveying continental literature. Dissections on monkey brains did not appear to support Owen's views. His friends at the University College Hospital in London dissected the brains of chimps, again confirming Huxley's suspicions. The soul of Man could not be found in any distinct feature 'which could be weighed, or measured, drawn or figured, calculated in inches and ounces'. The argument rumbled on at the BAAS in Oxford, and the following year in Manchester. Huxley was certain that Owen had made 'a prodigious blunder . . . he will be the laughing stock of all the continental anatomists'.

Then, after Owen's BAAS presentation in Cambridge in 1863, Huxley came down 'like the wolf on the fold', to the delight of the press, which revelled in parodying the fight. He proved there was nothing anatomically unique in the human brain. His research showed that not only did these three special anatomical features exist in apes, but they were sometimes even better developed than in man. He highlighted Owen's 'grave errors' and the 'utter baselessness' of his assertions, which had enabled this 'preposterous controversy' to drag on for two years.

An anonymous satire of their argument, *A Sad Case, Recently Tried before the Lord Mayor, Owen vs Huxley*, was published in 1863.

> Thomas Huxley, well known about town in connection with monkeys, and Richard Owen, in the old bone and bird stuffing line, were charged by Policeman X with causing a disturbance . . . Huxley was snapping his fingers at Owen and telling him he was only a little better than an ape . . . Huxley had got a beast of a monkey and said t'was his grandfather . . . he put the beast as near as ever he could to Owen and kept singing out 'Look at 'em, a'nt they like as peas?' . . . Owen behaved uncommon plucky, though his heart seemed broke. He tried to give Huxley as good as he gave, but he could not, and some people cried, 'Shame,' and 'he's had enough.' Never saw a man so mauled before. 'Twas the monkey that worrited him, and Huxley's crying out, 'There they are – bone for bone, tooth for tooth and their brains one as good as t'other.

In the same year Huxley's own book, *Man's Place in Nature*, was published. He summarised the evidence that Man, far from being a

Cartoon of Richard Owen riding a *Megatherium* skeleton.

special creation, could be placed in Nature just like the other beasts of the field. Ignoring Victorian repugnance at the idea, Huxley showed that Man's ancestors were the gorilla and the orang-utan. He took the opportunity to sneer at Owen's metaphysical language, such as his cherished, much repeated but undecipherable axiom: 'the continuous operation of the ordained becoming of organic forms'. 'It is obvious,' scoffed Huxley, 'that it is the first duty of a hypothesis to be intelligible, and this . . . may be read backwards, or forwards, or sideways, with exactly the same amount of signification.'

Hostility between the two men reached the point where Huxley used his power to oust Owen from key committees where he had reigned supreme for decades. In 1861, when Huxley was appointed on to the Zoological Society Council, Owen promptly stepped down. Within a year, Huxley took action to block Owen's move on to the Royal Society Council. Owen, for so long undisputed king of these establishment institutions, having used his power to blight promising careers such as Robert Grant's twenty years earlier, now found himself the hapless victim of similar manoeuvres. At the Royal Society, Huxley claimed the Council should not admit Owen, since he was 'guilty of wilful and deliberate falsehood'. The change in fortunes was so fast that, by 1862, Owen's stronghold at the Royal College fell to the evolutionists, as Huxley himself was honoured with Owen's former title: Hunterian Professor. 'I don't know what our illustrious predecessor will say,' scoffed Huxley's friends.

Owen's gilded reputation tarnished rapidly, his downfall inextricably linked with Darwin's rise. The *Bridgewater Treatises*, for so long his inspiration, were now lampooned as the *Bilge-water Treatises*. Owen was portrayed in *Vanity Fair* as 'Old Bones', and even described as a 'simple minded creature'. For *Punch* readers he was dismissed in a few verses:

> Next Huxley replies
> That Owen he lies
> And garbles his Latin quotation;

That his facts are not new
His mistakes not a few,
Detrimental to his reputation.

As Charles Darwin's ideas continued to gain momentum, soon Richard
Owen's own creation, the mighty dinosaurs, were wrested from his
care. Thomas Huxley, always on the attack, sought to place them
within an evolutionary framework. Were dinosaurs suddenly introduced upon the earth's surface, as Owen had written in 1842? In the
early 1860s, the weight of evidence supported this view. *Iguanodon* and
other stupendous monsters of the Mesozoic era seemed to appear from
nowhere in the fossil record. The great excitement prompted by the
discovery of reptiles in the Elgin sandstone had evaporated when the
rocks were correctly positioned in the sequence of strata. Far from
being Devonian in age, the Elgin rocks were formed in the later Triassic
period. So where were the primitive reptiles – the ancestors of the
dinosaurs?

In the early 1860s, the endless demand for fuel to feed the industrial
revolution drove miners deeper into the rich coal-seams of the north.
Fragments of strange primitive creatures began to emerge from this
blackened Carboniferous world of ancient swamps and giant forests.
Here were fossils like fish, but with stumpy legs and feet instead of fins.
These were the earliest amphibians, locked in their ironstone graves
around the coal formed an untold number of years ago.

At the School of Mines, Huxley studied the evidence. He was particularly interested in a group known as '*Labyrinthodonts*', which had
highly convoluted foldings in the enamel of their teeth, like a labyrinth
in cross-section. Their skeletal characteristics were unexpectedly like
reptiles. In 1861, only three genera of European Carboniferous
Labyrinthodonts were known. By 1865, Huxley was writing to Lyell to
inform him of some 'thirty genera of Labyrinthodonts known from all
parts of the world'.

It took time for a pattern to emerge, for the fish to be disentangled
from the amphibians. Some *Labyrinthodonts* were aquatic and could be as

large as alligators. Others had stout legs with a sprawling gait that would have enabled them to move well on land. Gradually, a picture began to form. The Devonian period, when fishes dominated ancient seas, had been followed by the Carboniferous period when creatures began to slither out of the water. With their short, sturdy legs, amphibians now took their first crawling steps on to the land.

Huxley did not stop there. If amphibians such as *Labyrinthodonts* were the predecessors of the great reptiles, what were their descendants? In the autumn of 1867, he visited the Ashmolean Museum to see Buckland's collection of *Megalosaurus* bones. Among the fossils, he was impressed by the *bird-like* structure of some of the dinosaur pelvic bones. This was beginning to fit evidence from America. A new dinosaur had been identified, named *Hadrosaurus*, which from the configuration of the skeleton was clearly bipedal. Even more bizarre, huge fossil footprints had been found, 18 inches long and resembling the tracks of giant birds, embedded in the Triassic rock. Was this 'Noah's raven' as reported in the press, or the footprints of early dinosaurs?

Huxley became intrigued with the bold idea that dinosaurs were, in some way, the predecessors of modern birds. He went back to the British Museum to view Mantell's collection of *Iguanodon* bones, and realised, as Mantell had argued, that the animal could be reconstructed as a bipedal creature: 'a sort of cross between a crocodile and a kangaroo, with a considerable touch of bird about the pelvis and legs,' he said. Huxley was helped in his analysis by the sensational discovery of a small carnivorous dinosaur, called *Compsognathus*, in the Jurassic rocks of Bavaria. Little more than two feet in length, this bipedal carnivore had marked bird-like features. 'Notwithstanding its small size,' he wrote, 'this reptile must, I think, be placed among or close to the Dinosauria; but it is still more bird-like than any of the animals which are ordinarily included in this group.'

He began to outline a bold classification showing how 'all living beings have been evolved one from the other'. There can be no doubt, Huxley observed, 'that the hind-quarters of the Dinosauria wonderfully approached birds in their general structure'. With remarkable insight he traced birds back, through large, flightless ancestors, to the dinosaurs of the Mesozoic era. 'The road from Reptiles to Birds is by way of Dinosauria,' he told a friend in January 1868. The next month, he presented his compelling ideas on 'Animals which are most nearly intermediate between Birds and Reptiles' before the Royal Institution. At last, dinosaurs were beginning to fit into the history of life on earth, with amphibian forebears and birds for descendants. They were no longer 'created' by God as fitting beasts for the primitive globe, 'rejoicing' in a complex reptilian form which had since degenerated. Owen's very reason for inventing the mighty *Dinosauria* was evaporating fast.

Although Owen did not accept Huxley's classification, he did not discount all the evidence for evolution. He tried to steer a path that allowed for a 'divinely plotted path of adaptive changes'. His objections centred on the idea that change was brought about randomly, by 'natural selection'. He aimed to avoid the opportunism of Nature and retain Divine intention; grand Divine laws ordered the material world. For Owen, dinosaurs were not an accident of nature, but he did accept that 'there was a certain systematic regularity in the order of their appearance'.

One by one, Owen's cherished notions on the dinosaurs were seen to fall apart. Ten years after Huxley lectured at the Royal Institution on their evolutionary history, a spectacular discovery in Belgium proved beyond doubt that Owen had wrongly interpreted the shape of the dinosaurs. In 1878, coal miners at Bernissart working more than one thousand feet underground realised they had drilled through some giant bones. Experts were summoned from the Musée Royal d'Histoire Naturelle de Belgique. Another tunnel was excavated one hundred feet lower than the first. This, too, contained fossil bones. They had uncovered the ghostly remains of the first dinosaur graveyard – a mass grave of *Iguanodons*.

The bones were excavated and taken to the Chapelle Saint-Georges in Brussels. Beneath the Gothic arches and stained-glass windows, they were reassembled: the femur, tibia, claw bones, the bones of the massive pelvis and shoulder, the vertebrae. As the first entire skeleton took shape, *Iguanodon* finally emerged from its hidden past. Gideon Mantell, fired by James Parkinson's words all those years ago, had always longed for just a few connected portions of skeleton. Twenty-five years after his death, the Belgian researchers produced thirty-one skeletons, the blueprint for which Mantell had sacrificed so much.

One glance at the skeletons confirmed the bipedality of *Iguanodon*. Far removed from Owen's four-footed reconstruction for the Crystal Palace, several features of the anatomy of the pelvis and the limbs confirmed the more upright posture suggested by his rivals, first Mantell and later Huxley. The specialist team, led by the naturalist Louis Dollo, could confirm many of Mantell's other inferences, such as his interpretation of the vertebrae and teeth and his suggestion that the forearms were used for seizing and grasping food.

Both Owen and Mantell had wrongly interpreted the great bony spike. This was not a nose horn, as in a rhinoceros, but formed the base of a huge spike on the thumb used for defence. The tail, too, could be used as a weapon, for striking any attacker. Dollo showed there was a lattice-like arrangement of tendons crossing the vertebrae of the tail, conferring both strength and rigidity. For decades, Mantell and Owen had had to guess the size of the beast by comparisons with other creatures. Dollo could simply measure the backbone and prove that the skeletons ranged from 13 to 30 feet in length. His thirty-year study of the creatures in the Belgian mass grave has made *Iguanodon* one of the best-known dinosaurs.

With new dinosaur discoveries, even Owen's classification became inadequate and out of date. When Owen created the dinosaurs, his classification was based on only three forms, *Megalosaurus*, *Iguanodon* and *Hylaeosaurus*, which were placed together as a single group. But in America, the end of the Civil War brought an era of rapid growth and exploration. Train lines were built that spanned the continent. Vast new fossil fields were found, heralding the spectacular dinosaur 'gold rush'.

Two leading American palaeontologists, Edward Cope and Othniel Marsh, named over one hundred and thirty new dinosaur species. An astonishing array of monsters began to surface from the Mesozoic world:

Diplodocus, Triceratops, Stegasaurus, Apatosaurus, Allosaurus, Ceratosaurus, Camptosaurus and *Brontosaurus*. Cope and Marsh revealed the immense variety of life during this era: plated dinosaurs like *Stegasaurus*, heavily armoured reptiles like *Triceratops* and *Nodosaurus*, horned dinosaurs, duck-bills such as *Trachodon*, new carnivores and herbivores. Many of their finds were complete skeletons, enabling a correct interpretation of their form.

Their new discoveries rapidly outstripped Owen's interpretations. Dinosaurs that Owen believed were aquatic were redefined as terrestrial. Some of the reptiles they found were far larger than anything Owen had anticipated. The thigh bone of Marsh's *Titanosaurus* alone was eight feet long; some monsters that they produced were sixty feet long. Both Cope and Marsh, like Huxley, attempted evolutionary classifications, as more evidence emerged that dinosaurs evolved from primitive amphibians. After visiting the Crystal Palace dinosaurs on a European tour in the 1890s, Othniel Marsh mocked Owen's famous recreations: 'so far as I can judge there is nothing like unto them in the heavens or on the earth, or in the waters under the earth'.

Richard Owen was left behind, an increasingly frail figure who lived to see his science overtaken by new ideas. He surfaced occasionally to lecture on dinosaurs or other extinct animals, but for the last few years of his life his main preoccupation was to last long enough to see the completion of his Natural History Museum in South Kensington. 'As my strength fails and I feel the term of my labours drawing nigh, how I long to see the conclusion of their main aim!' he told a friend. He nursed his bronchitis by the fireside, 'hoping still to survive to see the arrangement in systematic order of the national treasures of natural history in their noble new building'.

When the construction was complete in 1880, the stooped figure of Richard Owen, supported by his favourite curiously carved stick, could be seen fussing around the collections and supervising their display. As he mingled among the remains of the dinosaurs, fretting over details of glazes or position, he seemed to have become almost indistinguishable from the relics around him.

Richard Owen.

Intriguingly, a number of fossils from Gideon Mantell's collection never reached the galleries of the grand new Natural History Museum, but were sent to other museums as gifts or exchanges. Given what we know about Owen's character, it seems quite plausible that this dispersal of Mantell's treasured collection might have reflected the final phase of his victory over Mantell. Of the recorded transfers, some of these were presented to the Museum of Science and Art in Dublin, others to the Cheltenham College, the Seville Museum, Marlborough College and the Street Museum. Several Mantellian specimens mysteriously re-appeared in the collection of the Royal College of Surgeons. Since the fossils in the transfers were from the Tilgate Forest, even the *Iguanodon*, it is possible that Owen took some satisfaction in ensuring that no one could ever see all on one site the weight of evidence provided by Mantell that had helped to inspire his own ideas on the dinosaur.

Owen also triumphed in his battles for funds for the new museum. Backed by William Gladstone, this was indeed a cathedral to Nature. The central gallery was vast, with Gothic decoration, and domed ceilings large enough for the entire skeleton of a whale to be manoeuvred into the entrance hall, as he had always wished. But in the decades it took to build the museum, the natural sciences themselves had fundamentally changed, embracing a materialism that was shocking to Owen. As he took his place as Director, he was quite alone. His wife Caroline had died, after a protracted illness. The scientific community had moved on without him.

It took one strange and very personal event to finally shatter Owen's peace of mind. Shortly after the museum was opened to the public, his only child, William, committed suicide. On 13 March 1886, he went down to the Thames near his home in Mortlake, removed his hat and carefully placed his purse, watch and address card inside, and left it on the bank. His body was retrieved from the river by police the next day.

The tragedy was inexplicable to Owen. Whether William felt too intensely his father's 'lamentable coldness of heart', or was burdened by a job that held no interest for him and by the demands of a large family, the reasons remain unknown. This unexpected event, in addition to everything else, proved to be the final shock from which Owen never recovered. He retreated to his library at Sheen Lodge. His son's family moved in with him, but the ageing grandfather of Victorian science, shuffling around the top rooms of the house, became something of an object of fear to his grandchildren.

When Owen died, in 1892, he was 'systematically written out of history' by the Darwinians, according to the historian of science Nicolaas Rupke. Six hundred scientific papers and a lifetime's contribution to science were forgotten, as he was remembered principally for his opposition to Darwin. His personality was blackened, his treatment of rivals condemned, and the once brightest star in the scientific firmament faded from view. So complete was the assassination of his reputation that, within a few years, one Oxford professor dismissed him merely as 'a damned liar. He lied for God and for malice. A bad case.'

Epilogue

Vain the ambition of Kings
Who seek by trophies and like things
To leave a living name behind
And weave but nets to catch the wind.

Anonymous

All that remains of the struggles of the early pioneers are the fossils they retrieved from their buried world. Today, at the University Museum in Oxford, William Buckland's *Megalosaurus* jaw, so expertly interpreted by Georges Cuvier as of reptilian origin, is displayed beside the giant thigh bone, amid the fishes, amphibians and other primitive creatures that illustrate the march of evolution. After Buckland's lingering illness and death – the final irony for a man who believed the Creator gave every creature 'a dispensation of kindness to make the end of life to each individual as easy as possible' – a memorial was placed in his honour in Westminster Abbey, near the cloisters. 'He applied the Powers of his Mind to the Honour and Glory of God,' reads the inscription.

Many of Mary Anning's remarkable sea lizards line the walls of Gallery 30 in the Natural History Museum in London. The skull of Mary's first *Ichthyosaurus*, found by her brother Joseph below Black Ven in 1811, still survives, although it has been separated from the rest of the body. With a registration number painted under its huge bony eye, there is little to hint of the drama behind no. R.1158. This is the very fossil that provided £23 for the Annings when they were on poor relief and inspired the initial interest in the improbable creatures from 'former worlds' buried in the cliffs of England.

Mary Anning's first *Plesiosaurus* retrieved from the shore at Lyme in December 1823 is also mounted in this gallery: no. 22656. The apparent break at the base of its greatly elongated neck – which nearly cost Mary her livelihood when Georges Cuvier declared such an improbable creature could not exist – is plain to see. The fossil can be found opposite the museum restaurant, above the spot where parties of children sit to eat their packed lunches, unaware of the little piece of history mounted above their heads.

Gideon Mantell's Brighton collection, one of the first museums of giant land reptiles on which he pinned such great hopes of success, no longer exists. Originally twenty thousand fossils, some have been sold or lost, and many are archived underground, superseded by much more dramatic discoveries. The famous *Iguanodon* tooth, which Charles Lyell took to Cuvier in Paris, is now item MNZ GH 004839 in the Museum of New Zealand, Te Papa Tongarewa. Many other fossils which Mantell's son Walter took to New Zealand have lost their labels or been dispersed, the very outcome that Gideon Mantell most dreaded. Many were placed in the Colonial Museum, Wellington, and later transferred from one site to another as the Colonial Museum became first the Dominion Museum and later the National Museum.

In London, in the Natural History Museum, the 'Mantell-piece' unearthed from a quarry in Maidstone in 1834, which provided Mantell with the first connected portions of the *Iguanodon* skeleton, has been placed by the exit of Gallery 21. Opposite is a print of Mary and Gideon Mantell, and on a shelf below, perhaps symbolically placed between them, is an *Iguanodon* tooth, not unlike the very first that Mary found on the roadside. Still embedded in the rocks of the Weald that proved so hard to interpret, it is a poignant symbol of Mantell's painstaking struggle to understand a vanished world, a world so compelling that he sacrificed his marriage and his professional practice to this one bewitching interest. Now eclipsed in this gallery of wonders by the towering skeletons of the dinosaurs themselves, the tooth makes almost no claim on the attention of the jostling crowds.

Gideon Mantell's own spine, twisted into a macabre shape by his injuries, remained on display at the Royal College of Surgeons for almost a century and even inspired the odd scientific paper on the pathology of deformities of the backbone. In 1926 the specimen, no. 4808.1, was painstakingly remounted, described and catalogued. Some years later, during the Second World War, it was obliterated by German bombs at the height of the London blitz.

As for Richard Owen, his presence still haunts the Natural History Museum that he created in South Kensington. His imposing statue, unnoticed for the most part, commands the sweeping double staircase beyond the *Diplodocus* in the entrance hall. From this suitably pivotal position, his bronze eyes look down on a transformed world, one in which the vision of natural science is far removed from his own. The echoing stone floors and cathedral-like halls that he designed as a monument to God's wisdom and Divine natural laws now resonate to very different themes: gallery after gallery illustrates the evolutionary ideas of his rival.

It is the familiar, and enlarged, image of Darwin's face that hangs like a banner in the entrance, beneath the Gothic stained-glass windows. An entire floor is devoted to describing his ideas in the *Origin of Species*, which Owen reviewed so unfavourably. The natural world is laid out exclusively in terms of evolution – from the corals and sponges of primitive seas to the evolution of Man himself from ape-like primates. The dinosaurs themselves are no longer the embodiment and proof of the guiding hand of God, but an assemblage of strange monsters arising from a mere accident of nature.

Only ruined terraces with bramble-covered colonnades remain to hint of Joseph Paxton's former creation, which was razed to the ground by such a raging fire in 1936 that the flames could be seen over eight counties. A television transmitter, the Crystal Palace Stadium and a car park now dominate the crest of the hill.

As for Owen's rhinocerine dinosaur models that inspired the Victorian imagination in 1854, they can still be seen in the grounds of the Crystal Palace at Sydenham in South London. Once the proud trophies of a newly discovered science glimpsed through the splash of fountains in the gardens of the Crystal Palace, they have been stripped of their nineteenth-century glory. Chipped and broken, their paint long since faded, they seem strangely out of place: monstrous gargoyles peeping out at the twenty-first century from rampant undergrowth, a bizarre reminder of forgotten hopes and forgotten quarrels. Caught up in some uniquely British bureaucracy, they have been classified as Grade One listed buildings.

Notes and Sources

For details of those publications for which author's surname and short title are given here, see the Select Bibliography on p. 363.

CHAPTER I

Professor Hugh Torrens, geologist and historian of science at the Department of Earth Sciences, Keele University, is a leading expert on Mary Anning and has undertaken extensive searches of the available archives. He has summarised his research in 'Mary Anning of Lyme; the greatest fossilist the world ever knew', *British Journal of the History of Science*, vol. 28 (1995), pp. 257–84, and in an inspiring talk as keynote speaker at the Mary Anning Bicentennial Celebration in Lyme on 2–4 June 1999.

Many details of Mary Anning's background have also been gathered by the science historian William Lang (1878–1966). A comprehensive account of her life can be found in his 'Mary Anning of Lyme, Collector and Vendor of Fossils', *Natural History Magazine*, vol. 5, no. 34 (1936), pp. 64–81. In addition, Lang published many articles in the *Proceedings of the Dorset Natural History and Archaeological Society*. The papers cited in this chapter include 'Mary Anning and the Pioneer Geologists of Lyme', vol. 60 (1939); 'Three letters by Mary Anning', vol. 66 (1944); 'More about Mary Anning', vol. 71 (1949); 'Mary Anning and Anna Maria Pinney', vol. 76 (1956); 'Mary Anning's Escape from Lightning', vol. 80 (1959); 'Mary Anning and the Fire at Lyme', vol. 74 (1959);

'Portraits of Mary Anning', vol. 81 (1959); 'Mary Anning and a Very Small Boy', vol. 84 (1963).' These articles convey many aspects of her life and background.

The earliest reports of Mary Anning's work were published by the Lyme historian George Roberts, 'The fossil finder of Lyme Regis', *Chambers Journal of Popular Literature*, vol. 7 (1857), pp. 382–4. See also Charles Dickens's journal *All the Year Round*, vol. 13 (1965), pp. 60–3. The tragic story of the death of her older sister in a house fire appears in the *Bath Chronicle*, 27 Dec. 1798, p. 3. Additional material on her life and character has been written by Crispin Tickell, *Mary Anning of Lyme Regis*, published in 1996 by the Lyme Regis Philpot Museum. Information on the Poor Laws and the social history of the time can be found in G. M. Trevelyan, *English Social History* London/New York: Penguin, 1942.

'The London Baronet' Sir Everard Home's six papers attempting to describe Mary Anning's strange sea lizard first appeared in 1814 in 'Some account of the Fossil Remains of an Animal more nearly allied to Fishes than any other classes of Animals', *Philosophical Transactions of the Royal Society*, pp. 571–7. Home also published in the same journal: in 1816 pp. 318–61; 1818, pp. 24–32; 1819, pp. 209–11; 1819, pp. 212–16; 1820, pp. 159–64. The Reverend William Conybeare and Henry de la Beche published their first highly regarded paper, 'Notice of the Discovery of a New Fossil Animal', in *Transactions of the Geological Society*, vol. 5 (1821), pp. 559–94. More recently, J. B. Delair outlined these discoveries in 'A history of the early discoveries of the Liassic Ichthyosaurs', *Proceedings of the Dorset Natural History and Archaeological Society* (1968), pp. 115–27.

William Buckland's colourful early years have been portrayed by his children. His daughter Anna B. Gordon wrote *The Life and Correspondence of William Buckland* (London: John Murray, 1894). His son Francis Buckland wrote a 'Memoir of the Author' which appears in the 1858 edition of Buckland's *Bridgewater Treatise*. Nicolaas A. Rupke in *The Great Chain of History* (Oxford: Clarendon Press, 1983) has provided a detailed analysis of Buckland's contribution to English geology.

For the story of William Smith and the alleged plagiarism of his ideas by the Geological Society, see J. G. C. M. Fuller, *Strata Smith and his Stratigraphic Cross-sections, 1819* (Geological Society of London, 1995). Smith's difficulties are also analysed in H. S. Torrens, 'Patronage and Problems: Banks and the Earth Sciences', in R. E. R. Banks and others (eds), *Sir Joseph Banks: a Global Perspective* (Royal Botanic Gardens, Kew, 1994, pp. 49–75.

Early interpretations of fossils can be found in Martin J. S. Rudwick, *The Meaning of Fossils* (Chicago: University of Chicago Press, 1972); see his ch. 3 for a fascinating analysis of Cuvier's ideas. Cuvier's historic paper on extinction is 'Mémoire sur les espèces d'Éléphants tant vivantes que fossiles', *Magasin encyclopédique*, vol. 3 (1796), pp. 440–5. See also G. Cuvier, *Essay on the Theory of the Earth* (Edinburgh, 1813).

Additional summaries of folklore and religious beliefs can be found in K. P. Oakley, 'Folklore of Fossils', *Antiquity*, vol. 39 (1998), pp. 9–16; and in H. Torrens, 'Geology and the Natural Sciences', in Vanessa Brand (ed.), *Science and the Victorian Age* (1998).

The specialist sources cited in this chapter include J. A. Carr, *The Life and Times of James Ussher, Archbishop of Armagh* (London: Wells, Gardner, Darton & Co., 1895). Edmond Halley described his tests of the age of the earth in 'A Short Account of the Cause of the Saltiness of the Ocean', *Philosophical Transactions of the Royal Society*, vol. 29 (1714), pp. 296–300. Horace Woodward in *The History of the Geological Society of London* (Geological Society of London, 1907) describes key characters and their aims. The conflict with religion is summarised in 'The Scriptural Geologists' by Milton Millhauser, *Osiris*, vol. 11 (1954), pp. 65–86, and *Genesis and Geology* by Charles Coulston Gillispie (Cambridge, Mass.: Harvard University Press, 1951).

CHAPTER 2

The first biography of Gideon Mantell was written by Sidney Spokes and published in 1927: *Gideon Algernon Mantell, LLD, FRCS, FRS, Surgeon and*

Geologist (London: John Bale & Sons & Danielson). Spokes, who lived in Mantell's house in Castle Place, gathered together a wealth of personal information about his background, letters and papers, and anecdotes such as the story of the ammonite and Mantell's meeting with James Parkinson. More recently, Professor Dennis Dean has published *Gideon Mantell and the Discovery of Dinosaurs* (Cambridge University Press, 1999). After an exhaustive search of all the unpublished Mantell archives including those in New Zealand, Dean provides a scholarly account of Mantell's life and achievements.

Other references to Mantell's early years given here can be found in his obituary, 'Gideon Algernon Mantell, 1790–1852', *The Gentleman's Magazine* (Dec. 1852), which gives an account of his education and dissenting background. A. D. Morris, 'Gideon Algernon Mantell, Surgeon, Geologist and Wizard of the Weald', *Proceedings of the Royal Society of Medicine*, vol. 65 (1972), pp. 215–21, provides information on his role as a doctor. R. J. Cleevely and S. D. Chapman, in 'The accumulation and disposal of Gideon Mantell's fossil collection', *Archives of Natural History*, vol. 19, no. 3 (1992), pp. 307–64, discuss his social aspirations. Mantell described his life as a doctor in *Memoirs of the Life of a Country Surgeon* (London: 1848). Additional material on his early years can be found in J. B. Delair and Dennis Dean, 'Gideon Mantell in Wiltshire', *Archaeological and Natural History Magazine*, vol. 79 (1985), pp. 219–24, and W. E. Swinton, 'Gideon Algernon Mantell' *British Medical Journal* (1975), pp. 505–8.

The most vivid insights into his daily life and his ambitions come from Mantell's own journal. Four volumes of his unpublished diaries, from 1819 until his death in 1852, are now stored at the Sussex Archaeological Society in Lewes, Sussex. There is an edited version of his diary by E. C. Curwen, *The Journal of Gideon Mantell, Surgeon and Geologist: 1818–1852* (London: Oxford University Press, 1940). Mantell's vision of the role of the geologist is quoted from his very popular book, *Thoughts on a Pebble* (London: Reeve, Benham & Reeve, 1849).

The life of the campaigner Thomas Paine, who influenced Mantell's father, is discussed in C. Brent, *Georgian Lewes 1714–1830: The Heyday of a Country Town* (Colin Brent Books, 1993). New research has shown that Mantell's father was also a 'Master Gardener' of some repute (Hugh Torrens, personal correspondence). The early history of the Royal Society is outlined in Weld, *History of the Royal Society* (London: 1848).

The quotations from James Parkinson included in this chapter are from his book *The Organic Remains of a Former World; An Examination of the Mineralised Remains of the Vegetables and Animals of the Ante-diluvial World* (London: Robson White & Murray), published in three volumes between 1804 and 1811. Further discussion of Parkinson's contribution to English geology can be found in R. J. Cleevely and J. Cooper, 'James Parkinson, a significant English eighteenth-century Doctor and Fossil Collector', *Tertiary Research*, vol. 8, no. 4 (1987), pp. 133–44; and in M. Critchley, 'James Parkinson, a Bicentenary volume of papers dealing with Parkinson's disease . . .' (London and New York: Macmillan, 1955).

For information on some of Mantell's correspondents, see Ron Cleevely on Etheldred Benett in 'The First Female Palaeontologist', *The Linnean*, vol. 14, no. 2 (1998), pp. 3–9. Thomas Birch's contribution is described by H. Torrens, 'Colonel Birch', in *Collections and Collectors of Note*. Mantell's debt to the earlier work of John Farey, who was first to pioneer stratigraphic studies in Sussex, is discussed in H. Torrens, 'Coal Hunting at Bexhill, 1805–11: how the New Science of Stratigraphy was ignored', *Sussex Archaeological Collections*, vol. 136 (1998).

Few records of Mary Mantell survive, and most of the entries in Mantell's diaries which shed light on their domestic life have been deleted. Some details of his early relationship with her can be found in Dean, *Gideon Mantell*, pp. 28–31. Their meeting and George Woodhouse's medical treatment are also described in *The Gentleman's Magazine* (Dec. 1852).

For the sequence of events leading to the discovery of the *Iguanodon* I am indebted to Dr Angela Milner, Head of Fossil Vertebrates, and Sandra Chapman, Vertebrate Curator, in the Department of Palaeontology at the Natural History Museum, London, for many helpful discussions on the chronology and the evidence available to Mantell. Ron Cleevely, Scientific Associate in the Department of

Palaeontology at the Natural History Museum, and Professor Hugh Torrens of Keele University also provided invaluable information.

For details of the first fossil 'palm', see Joan Watson and Caroline Sincock, *Bennettitales of the English Wealden* (London: Palaeontographical Society, 1992); the trunk, discovered in 1820, is discussed on p. 186. Mantell's first book, *The Fossils of the South Downs*, published in London by Lupton Relfe in 1822, and his journal entries, give key insights into the fossils he acquired early in his career and his interpretations of them. Dennis Dean has also studied Mantell's important contribution to the discovery of dinosaurs and gives his fascinating analysis in 'Gideon Mantell and the Discovery of *Iguanodon*', *Modern Geology*, vol. 18 (1993), pp. 209–19, and more recently in Dean, *Gideon Mantell*, pp. 52–86.

There are several different versions of the discovery of the first *Iguanodon* tooth. Dennis Dean has suggested it is possible that Mr Leney the quarryman found it, perhaps as early as 1819. Edwin Colbert in *Men and Dinosaurs* (Penguin, 1968), Sidney Spokes, and W. E. Swinton in 'Gideon Mantell and the Maidstone Iguanodon' in *Notes and Records of the Royal Society of London*, vol. 8 (1951), pp. 261–76, have all endorsed the version in which Mary Mantell finds the first *Iguanodon* tooth, as here. Mantell publicly attributed the first find to his wife in both 1827 and 1833, although in later publications, after he had separated from her, he credited the first find to himself. The dates Mantell gives to the discovery are also unproved. Spokes, *Gideon Algernon Mantell*, identifies the find to the summer of 1822, but this is not possible since the *Iguanodon* teeth are described in Mantell's *Fossils of the South Downs*, which was completed several months previously. It seems most plausible from the available evidence that the herbivorous teeth were found during 1820 and 1821.

CHAPTER 3

The personal anecdotes about William Buckland's character cited in this chapter are from many different records and archives. The poem describing his rooms at Oxford can be found in Gordon, *Life and*

Correspondence, p. 9; some of the reminiscences of colleagues and students concerning his lectures and his hospitality also appear in this biography by his daughter. Francis Buckland, in a memoir introducing the 1858 edition of Buckland's *Bridgewater Treatise*, covers much of the same material, although in less detail. The story of Tiglath the bear can be found in several sources, but is most fully outlined in Edwin Colbert's *Men and Dinosaurs* (New York: Penguin, 1968), p. 24. An excellent overview of Buckland's early contribution to geology and the response to his work by biblical literalists is presented by Nicolaas A. Rupke in *The Great Chain of History* (Oxford: Clarendon Press, 1983).

Many original works by William Buckland are cited in this chapter. His inaugural lecture at Oxford, entitled '*Vindiciae Geologicae*, or The Connexion between Geology and Religion explained', was delivered on 15 May 1819 and published in 1820 in Oxford. Buckland's paper on the 'Description of the quartz rock of Lickey Hill in Worcestershire . . . with considerations on the evidence of a recent Deluge . . .', in which he describes his work on the Flood, appeared in *Transactions of the Geological Society of London*, vol. 5 (1821), pp. 506–44. His study of hyenas was published first by the Royal Society: 'An account of an assemblage of Fossil Teeth and Bones of Elephant, Rhinoceros, Hippopotamus, Bear, Tiger, and Hyaena . . . .', *Philosophical Transactions of the Royal Society* (Feb. 1822), pp. 171–230. A fuller version of this paper was published the following year in London, entitled '*Reliquiae Diluvianae*, or Observations on the organic remains contained in caves, fissures and diluvial gravel . . . attesting the action of a universal deluge'. The first edition of this sold out, and there was a second in 1824.

Unfortunately, William Buckland kept no records of the giant bones he found in quarries in Oxfordshire, or of his early interpretations. Correspondence between Buckland and several others, including Joseph Pentland, reveals that Cuvier visited the Ashmolean Museum in 1818 and concluded that the bones belonged to a reptile. I am indebted to Dr Angela Milner for helpful guidance on Cuvier's likely reasoning on the basis of the available fossil specimens and knowledge of reptilian anatomy at the time. See also William Buckland's paper, 'Notice on the

Megalosaurus or Great Fossil Lizard of Stonesfield', *Transactions of the Geological Society of London* (1824), pp. 390–7, which describes the strata and Cuvier's conclusions on the size of the beast, and shows which fossil bones had been uncovered.

The scriptural geologists quoted in this chapter wrote the following papers: Reverend George Young, *Scriptural Geology, or an Essay on the High Antiquity ascribed to the Organic Remains imbedded in Stratified Rocks* (London: Simpkin Marshall, 1840), p. 8; and also 'On the Fossil Remains of Quadrupeds' 1822, *Mem Wernerian Natural History Society*, vol. vi (1832); George Fairholme, 'A Layman on Scriptural Geology', *Christian Observer*, 1834, pp. 479–92; and George Bugg, *Scriptural Geology*, or *Geological Phenomena consistent only with the literal interpretation of the Sacred Scriptures*, 2 vols (London: Hatchard & Son, 1826–7). The view that strata could be formed instantaneously is presented in George Cumberland's 'Strata Formation . . . ', *Monthly Magazine*, vol. 52 (1821), pp. 301–5; in Reverend George Young (above); and in H. Torrens, 'Geology and the Natural Sciences . . .', in Vanessa Brand (ed.), *Science and the Victorian Age* (1998). A good summary of the issue is to be found in Milton Millhauser, 'The Scriptural Geologists', *Osiris*, vol. 11 (1954), pp. 65-86. Alternatively, there were scholars who questioned the wisdom of even attempting to reconcile the Bible and geology; see W. H. Fitton, '*Reliquiae Diluvianae* or Observations on Organic Remains', vol. 39 (1823), pp. 196–234.

Early discoveries of giant bones are discussed in J. B. Delair and W. A. S. Sarjeant, 'The Earliest Discoveries of Dinosaurs', *ISIS*, vol. 66 (1975), pp. 5–25. Robert Plot describes the first discoveries of giant bones in *The Natural History of Oxfordshire* (1677). Discussions of Buckland's ideas on the Flood, Lamarck and Cuvier can be found in the following volumes: Gillispie, *Genesis and Geology*; Sir Archibald Geikie, *The Founders of Geology* (London: Macmillan, 1897); and Rudwick, *Meaning of Fossils*. There is a translation of Lamarck's original writings by D. R. Newth, 'Lamarck in 1800: A lecture on invertebrate animals', *Annals of Science*, vol. 8 (1952), pp. 229–54.

For details of Georges Cuvier's interest in publishing details of the

Stonesfield fossil reptile, see 'An Irish Naturalist in Cuvier's Laboratory, The letters of Joseph Pentland, 1820 to 1832', *Bulletin of the British Museum of Natural History*, vol. 6, no. 7 (1998), pp. 245–319; letters dated 20 Sept. 1821, 25 Feb. 1822, 28 Feb. 1824. This correspondence also reveals Cuvier's interest in the hyena caves. See also Reverend William Conybeare's and Henry de la Beche's 'Notice of the Discovery of a new Fossil Animal forming a link between the Ichthyosaurus and the Crocodile . . .', *Transactions of the Geological Society*, vol. 5 (1821), pp. 559–94, the first English paper in which the Stonesfield lizard was mentioned.

CHAPTER 4

Mantell's interpretations of fossil plants cited in this chapter are from several original sources. As early as 1818, he had identified tropical plants that he thought similar to those of the 'Cactus tribe', and published his findings in 'A Sketch of the Geological Structure of the South Eastern part of Sussex', the *Provincial Magazine* (Aug. 1818), pp. 8–11. More details of plants that he thought resembled tree-ferns such as *Dicksonia*, and palms, are described in Mantell's *Fossils of the South Downs*, pp. 42 5, 57. In addition, see Mantell's letter to the Geological Society of 14 June 1822 in which he identifies fossil cycads and compares them to *Cycas revoluta* at Loddiges' Greenhouses; his paper is called 'On the Iron sand Formation of Sussex' and appears as the 'Notices and Extracts from the Minute Book of the Geological Society', published in *Transactions of the Geological Society of London*, vol. 2 (1826). Also in the 'Notices and Extracts . . .' Mantell identifies other plants in addition to those above, in 'Description of some fossil Vegetables in the Tilgate Forest in Sussex' (1823), pp. 421–4. His diary, too, records the discovery of some of these fossil plants; specimens that he thought resembled tropical *Euphorbiae* are identified from 1820. Many of these specimens were almost certainly *Bennettitales*, an extinct group of cycad-like plants dominant in the Wealden flora. I am indebted to Dr Joan

Watson of the Department of Geology, Manchester University, for advice on Mantell's early contributions.

The account of Mantell's first meeting with Charles Lyell cited here is from Mantell's correspondence with Professor Benjamin Silliman in 1841; see Spokes, *Gideon Algernon Mantell*. For the powerful influence of William Buckland on Charles Lyell early in his career, see L. G. Wilson, *Charles Lyell, The Revolution in Geology* (New Haven, Conn.: Yale University Press, 1972), p. 43 onwards. Lyell's background is also described in James Secord's introduction to *Principles of Geology* (London: Penguin, 1997 edn). Mantell's diary mentions meetings and correspondence, including an early exchange of Stonesfield fossils (see entries for Oct. and Nov. 1821).

The subscribers to Mantell's *Fossils of the South Downs* are acknowledged in its introduction. The section on the 'Limestone of Tilgate Forest' is on pp. 37–60; vegetable remains are on pp. 42–5; fossil shells, p. 45; fossil *Lacertae* (lizards), pp. 48–54; unknown animals, pp. 54–5; comparisons to the Stonesfield beds, pp. 59–60. Mantell's conclusion on gigantic animals of 'the Lizard Tribe' is on pp. 56, 304.

The meeting at the Geological Society when Mantell's herbivorous reptile teeth were wrongly identified by William Buckland, Conybeare and Clift as belonging to a wolf-fish or a large mammal, is described in numerous sources. This account is derived from Mantell's own recollections outlined in Spokes, *Gideon Algernon Mantell*, and also in Mantell, *Petrifactions and their Teachings* (London: Bohn, 1851), p. 229. This meeting is likely to have occurred after the publication of *Fossils of the South Downs* in May 1822 but before Mantell's letter to the Geological Society of 1 June of that year, in which he refers to the specimens from Tilgate shown to the Society; the most likely date is 17 May, when the minutes of meetings at the Society show that Mantell discussed the Tilgate specimens. His difficulties in identifying the strata of the Weald are described in *Fossils of the South Downs*, pp. 57–9, 295–303.

Regarding prejudice against provincials as experts, this was observed by the geologist Robert Bakewell and quoted in H. S. Torrens, 'The scientific ancestry and historiography of the Silurian System', the

Quarterly Journal of the Geological Society, vol. 147 (1990), pp. 657–62. Torrens's article, p. 659, also describes Greenough's 'highly political grip' on the Geological Society. The quotation from William Smith comes from Horace Woodward, *The History of the Geological Society of London* (London: Geological Society, 1907).

The evidence that by 1822 Mantell had correctly identified the Tilgate strata as part of the iron-sand formation and that the herbivorous teeth belonged to an unknown reptile can be found in his letter to the Geological Society of 1 June 1822. Minutes of the Society, pp. 340–3, show that this was read as a paper from 'Mr Mantell and Mr Lyell on the Iron-sand of Sussex' on 17 Jan 1823. It was eventually published as 'On the Iron sand Formation of Sussex' and appears as the 'Notices and Extracts from the Minute Book of the Geological Society', published in *Transactions of the Geological Society of London*, vol. 2, part 1 (1826), pp. 130–4. Although, with a backlog of papers, some delay between reading a report and its subsequent publication in *Transactions of the Geological Society of London* is to be expected, it is noticeable that writers such as Conybeare, de la Beche, Buckland and Murchison were published more quickly than Mantell. Mantell's difficulties with referees can be found in the Geological Society's Referees' Reports, 1818–25, Com/p. 4/47 (Greenough) and Com/p. 4/49. For Buckland's warning Mantell against publication, see Spokes, *Gideon Algernon Mantell*, p. 21.

I am indebted to Dr Ron Cleevely and Professor Hugh Torrens for detailed discussions on the freshwater nature of the Weald and the contribution made by William Fitton. There appears to have been some ill-feeling between Fitton and Mantell as to who deserved the credit for this work. Publicly, Charles Lyell credited his friend Gideon Mantell several times. Their correspondence, too, reveals that they were discussing freshwater invertebrates from 1822, and some of these are also identified in Mantell, *Fossils of the South Downs*, p. 304. The *Dictionary of Scientific Biography* entries on both Lyell and Mantell also credit Mantell. However, other research shows that obtaining exact proof took several years. Fitton himself published his findings between 1824 and 1836, and claimed to have elucidated the freshwater nature of the Weald himself; see M. A. Challinor, 'The Beginnings of Scientific Palaeontology in Britain', *Annals of Science*, vol. 6, no. 1 (Oct. 1948). More information on William Fitton's contribution to early geology can be found in Horace Woodward, *The History of the Geological Society of London* (London: Geological Society, 1907).

Lyell's letter correctly confirming the Weald as Secondary rock and comparing it to strata in the Isle of Wight is in Katherine M. Lyell (ed.), *Life, Letters and Journals of Sir Charles Lyell Bart.*, 2 vols (London: John Murray, 1881); see pp. 121–2. Background information on Cuvier's soirées etc. comes from 'An Irish Naturalist in Cuvier's Laboratory, The letters of Joseph Pentland, 1820 to 1832', *Bulletin of the British Museum of Natural History*, vol. 6, no. 7, pp. 245–319. Cuvier's interpretation of the herbivorous teeth is outlined in Spokes, *Gideon Algernon Mantell*, and Mantell, *Petrifactions and their Teachings* (London: Bohn, 1851). His re-interpretation the next morning is described in J. C. Yaldwin, G. J. Tee and A. P. Mason, 'The status of Gideon Mantell's first *Iguanodon* Tooth in the Museum of New Zealand, Te Papa Tongarewa', *Archives of Natural History*, vol. 24 (1997), pp. 397–422.

There is considerable evidence in Mantell's diary and correspondence that domestic conflict between him and his wife was beginning to surface by 1822, with several references to his unhappiness and frustration. The details of the finances of *Fossils of the South Downs* and the contribution of Mary Mantell's brother are discussed in Dean, *Gideon Mantell*, p. 51, footnote.

CHAPTER 5

Mary Anning's discoveries of ichthyosaurs are summarised in J. B. Delair, 'A history of early discoveries of Liassic Ichthyosaurs in Dorset and Somerset', *Proceedings of the Dorset Natural History and Archaeological Society* (1968), pp. 115–27. Details of her devoted dog, Tray, are provided in W. D. Lang, 'More about Mary Anning, including a Newly Found Letter', *Proceedings of the Dorset Natural History and Archaeological*

Society, vol. 71 (1949), pp. 184–8. The characteristics of the *Plesiosaurus* that she uncovered on the evening of 10 December 1823 are outlined in the Reverend Conybeare's paper describing the new animal: 'On the Discovery of an almost perfect Skeleton of the Plesiosaurus', *Transactions of the Geological Society of London*, vol. 1 (1824). This paper also provides a detailed description of the anatomy of the *Plesiosaurus* and speculations on its habitat.

I am indebted to Philippe Taquet at the Muséum National d'Histoire Naturelle in Paris for information on Georges Cuvier's initial response to the discovery of the *Plesiosaurus*. Conybeare's early papers outlining his reasons for believing that such a creature might exist are in two papers in *Transactions of the Geological Society of London*: 'Notice of the discovery of a new Fossil Animal, forming a link between the Ichthyosaurus and the Crocodile, together with general remarks on the Osteology of the Ichthyosaurus' (1821), and 'Additional Notices on the Fossil Genera Ichthyosaurus and Plesiosaurus' (1822).

An account of the eccentric collector Thomas Hawkins, who sometimes embellished fossils, can be found in a paper by W. D. Lang, 'Three letters by Mary Anning', *Proceedings of the Dorset Natural History and Archaeological Society*, vol. 66 (1944), p. 171. Additional background information on the response of the geological community to Mary Anning's *Plesiosaur* is in George Cumberland Papers at the British Library, Add. MSS 36491–36522, vols for 1823 (36509) and 1824 (36510). Comments from Cumberland to his friend regarding the 'new fish' are in a letter dated 4 Jan. 1824, f. 1; Cumberland Papers, 36510. A letter from Charles Konig showing that he believed that the new animal was genuine is also in the Cumberland Papers, 36510, f. 31. Concerns that the creature could only be seen by candlelight in a passageway are expressed in 36510, f. 33.

Reverend Conybeare's excitement at the new discovery is portrayed in his letter to Henry de la Beche, which also reveals how the creature was delayed in the Channel and could not be brought into the Geological Society. This was first outlined in W. D. Lang, 'Mary Anning and the Pioneer Geologists of Lyme', *Proceedings of the Dorset Natural History and*

Archaeological Society, vol. 60 (1939), pp. 152–3. Geological Society minutes shed more light on the sequence of events leading to the announcements of the *Plesiosaurus* and *Megalosaurus*; see meetings for 6 Feb. 1824 and 20 Feb. 1824, pp. 404–12. Conybeare's eccentric lecturing style is described in H. S. Torrens, 'The Dinosaurs and Dinomania over 150 years', *Modern Geology*, vol. 18 (1993), p. 2. Buckland's view of the *Plesiosaurus* as the most 'monstrous' creature found 'amid the wreckage of the former world' is from the *Bridgewater Treatises*, 1836, vol. 1, ch. XIV; for the section on the *Plesiosaurus* see p. 202 onwards.

William Buckland's paper on the first named dinosaur is 'Notice of the Megalosaurus or Great Fossil Lizard of Stonesfield', *Transactions of the Geological Society of London* (1824), pp. 390–96; see also Geological Society minutes for 20 Feb. 1824. I am indebted to Professor Hugh Torrens for insights into the sequence of events leading to the landmark meeting where *Plesiosaurus* and *Megalosaurus* were first described, and for information regarding Buckland's correspondence with Cuvier in which Buckland expresses his desire to publish on *Megalosaurus* soon after press reports of Mantell's discoveries in Sussex. Professor Torrens drew my attention to the revealing unpublished letter between Warburton (on the publications committee of the Geological Society) and Buckland of 12 Mar. 1824, showing that Buckland hoped to incorporate some of Mantell's Sussex discoveries in his own paper. This is in the Devon Record Office, ref: 138m/f. 71, and I am grateful to Rosemary Gordon for kind permission to cite this letter.

The quotations on Mary Anning that shed more light on how her character developed as she became better known are from W. D. Lang, 'Mary Anning and the Pioneer Geologists', and Lang, 'Mary Anning of Lyme', *Natural History Magazine*, vol. 5 (1936), pp. 64–81; also from *Natural History Magazine*, Lang, 'More about Mary Anning, including a newly found letter', vol. 71 (1949), p. 187, and 'Mary Anning and Anna Maria Pinney', vol. 76 (1956), p. 147.

The progress of Gideon Mantell's work of identifying the fossil reptiles in his collection is revealed in Buckland's 1824 paper on the

Megalosaurus, which describes the *Megalosaurus* fossils in Mantell's possession and his research comparing the Stonesfield and Tilgate fossils. His efforts to identify the giant bones in his collection are also outlined in Mantell's own paper on the *Iguanodon*, 'Notice on the Iguanodon, a Newly discovered Fossil Reptile from the sandstone of the Tilgate Forest in Sussex', *Philosophical Transactions of the Royal Society*, vol. 115 (1825), pp. 179–86.

Georges Cuvier's reply to Mantell in June 1824 is quoted in many sources; see Spokes, *Gideon Algernon Mantell*, pp. 19–20, Dean, *Gideon Mantell*, p. 81, and G. A. Mantell, *Petrifactions and their Teachings* (London: Bohn, 1851), p. 231. Cuvier's letter in the original French is cited at length in Mantell's paper on the *Iguanodon* (above). This paper also describes the contribution made by William Clift at the Hunterian Museum and the similarities to the iguana. Mantell's efforts to gather a series of *Iguanodon* teeth are described in Spokes, *Gideon Algernon Mantell*, p. 21, and in G. A. Mantell, *Illustrations of the Geology of Sussex* (London: Lupton Relfe, 1827).

Letters from Buckland and Mantell to Cuvier in spring/summer 1824 are cited in P. Taquet, 'Georges Cuvier, Buckland et Mantell et les Dinosaures', *Symposium Paléontologique*, Montbéliard, France, 1982, available from the Muséum National d'Histoire Naturelle in Paris. The letter cited from Reverend Conybeare to Mantell on the name 'Iguanasaurus' is discussed in Dean, *Gideon Mantell*, p. 85. For Conybeare's 'sneering' style, which evidently offended several aspiring geologists, see Cumberland Papers, British Library, Add. MSS 36510, f. 4.

The anecdotes surrounding William Buckland's meeting with Mary Morland and her character are best described in his children's biographies. See Gordon, *Life and Correspondence*, and Francis Buckland's memoir introducing the 1858 edition of Buckland's *Bridgewater Treatises*. New evidence set out in the *Dictionary of National Biography* entries for William and Mary Buckland has cast doubt on the story of their meeting as described in family archives, by suggesting that Miss Morland had been known to William Buckland since 1820.

The events at the Royal Society described in this chapter are outlined in the Royal Society archive minutes dated 1822–6; see minutes of 10 Feb. 1825 for the reading of the paper on *Iguanodon*, and of 22 Dec. 1825 for Mantell's entry as a Fellow. See also Mantell's diary for these dates.

CHAPTER 6

There are numerous sources outlining details of Richard Owen's background. Many anecdotes are described in his grandson's biography: Richard Owen, *The Life of Richard Owen*, vol. 1 (London: John Murray, 1894), vol. 1, although the progress of Owen's scientific achievements is not well documented in this account. Fascinating insights into his character and recollections of his contemporaries are portrayed in Adrian Desmond, *Archetypes and Ancestors* (London: Blond and Briggs, 1982). Owen's contribution to science is discussed in N. Rupke, *Richard Owen, Victorian Naturalist* (New Haven & London: Yale University Press, 1994). His work is also discussed in K. Padian, 'The Rehabilitation of Sir Richard Owen', *BioScience*, vol. 47, no. 7 (1997). See also Dale Lloyd Ross, 'A survey of some aspects of the life and work of Sir Richard Owen', PhD thesis, 1972, University of London, Natural History Museum, Owen Collection, 73.

Richard Owen's early fears of the supernatural and the details of the ghost stories can be found in *Hood's Magazine and Comic Miscellany*, vol. 2 (1844), pp. 442–50, which published the story of the ghosts in the tower; and vol. 3 (1845), pp. 294–303, which relates the tale of the severed head. His 'confessions', for so long held secret, were submitted as letters to Mr Gideon Shaddoe in 1844, written by 'your confiding friend, Silas Seer'.

The effects of university life on Richard Owen are discussed in Owen, *Richard Owen*; see also J. D. Comrie, *History of Scottish Medicine*, vol. 2 (London, 1932). Alexander Monro the Third also had a major effect on Charles Darwin, who studied at Edinburgh a few years later. Darwin's impressions are given in Adrian Desmond and James Moore, *Darwin* (London: Penguin, 1992), p. 26 onwards.

The arguments of the early evolutionists Jean-Baptiste Lamarck and Geoffroy Saint-Hilaire are outlined in numerous sources. An excellent account can be found in Rudwick, *Meaning of Fossils*, pp. 115–20. Lamarck's work as one of the 'glories of French Science' is described in Geikie, *Founders of Geology*; see p. 350. Theories on the 'Chain of Being' are outlined in N. A. Rupke, *The Great Chain of History* (Oxford: Clarendon Press, 1983), p. 169 onwards. See also C. J. Schneer, *Towards a History of Geology* (Boston, Mass.: MIT Press, 1967), pp. 36–62, where Frank Bourdier discusses the antagonism between Geoffroy Saint-Hilaire and Cuvier.

Original citations given in this section on the Chain of Being include D. R. Newth, 'Lamarck in 1800: a lecture on Invertebrate Animals . . .', *Annals of Science*, vol. 8 (1952), pp. 229–54. Georges Cuvier's first works on woodlice are in *Journal d'histoire naturelle* (1792), Paris. William Buckland's thoughts on crocodiles are cited in Young, 'Account of a Fossil Crocodile', *Edinburgh Philosophical Journal*, vol. XIII (1825), pp. 76–81. *The Politics of Evolution* by A. Desmond (Chicago and London: University of Chicago Press, 1989) provides a fascinating perspective on evolution and social reforms; see chs 1 and 2. See also E. Royle, *Victorian Infidels: the Origin of the British Secularist Movement* (Manchester University Press, 1974), and L. S. Jacyna, 'Medical Science and Moral Science', *British Journal of the History of Science*, vol. 25 (1987), pp. 111–46.

Reptiles as links in the Chain of Being are discussed in several of the sources above. See also H. S. Torrens and M. A. Taylor, 'Saleswoman to a new Science' in *Proceedings of the Dorset History and Archaeological Society*, vol. 108 (1987), p. 142. The quotation from Lyell on the leap in the chain is given in a letter to Mantell 17 Feb. 1824; see Lyell, *Life, Letters and Journals*, vol. 1, p. 151. The memorial brass dedicated to John Hunter is on the north aisle in Westminster Abbey, inscribed in 1859 and quoted in John Hunter, *Dictionary of Scientific Biography*, pp. 566–8.

Personal accounts of Owen's first experiences in London and the significance of Abernethy are described in Owen, *Richard Owen*, vol. 1, p. 30. The difficulties facing the Royal College in the 1820s are shown in Desmond, *Politics of Evolution*, pp. 240–7. I have leaned heavily on Desmond's account of the Royal College in crisis, with senior figures at the College increasingly a target for the radical medical press. See also J. Dobson (1954), p. 50 onwards. For an alternative perspective on Everard Home's 'theft', see J. M. Oppenheimer, *New Aspects of John and William Hunter* (London: Heinemann, 1946), pp. 69–73.

The interest that Catherine Owen took in her son's career is found in J. W. Gruber and J. C. Thackray, *Richard Owen Commemoration* (London: Natural History Museum Publications, 1992), p. 71; this publication gives many details of his character and motivation. For Catherine Owen's comments on Cuvier see Owen, *Richard Owen*, p. 59; for the significance of allying himself with eminent men, see pp. 34–6; and for Richard Owen's meeting with Caroline Clift, p. 34. For further information on Cuvier's character, such as his annoyance with those who could not speak French, see 'An Irish Naturalist in Cuvier's Laboratory: the letters of Joseph Pentland, 1820 to 1832', *Bulletin of the British Museum of Natural History*, vol. 6, no. 7, pp. 245–319, introduction.

CHAPTER 7

William Buckland's paper on coprolites cited in this chapter is 'On the Discovery of Fossil Faeces, in the Lias at Lyme Regis and in other Formations', *Transactions of the Geological Society of London*, 2nd series, part 3 (1835), pp. 223–36. His studies are also discussed in his daughter's biography: Gordon, *Life and Correspondence*, p. 114; for the rhyme about the Flood, p. 26. A full analysis of the significance of Buckland's research on coprolites is in Rupke, *Chain of History*. I am indebted to this account for the verses on coprology written by his students at Oxford, cited on p. 142. Thomas Hawkins's melodramatic view of the ancient world can be read in *The Book of Great Sea Dragons* (London: William Pickering, 1840).

Details of Adolphe Brongniart's study of fossil plants were published in 1828 in his *Prodrome d'une histoire des Végétaux Fossils* (Paris). Mantell's work with Brongniart on horsetails is discussed in J. Watson and D. J. Batten, 'A revision of the English Wealden Flora 11, *Equisetales*', *Bulletin of the British Museum of Natural History*, vol. 46, no. 1 (May, 1990), pp. 37–60.

Gideon Mantell's work on the fossils of the Tilgate Forest and the progress of his research during 1826 and 1827 appears in his short book, *Illustrations of the Geology of Sussex* (London: Lupton Relfe, 1827); see especially ch. 2, 'A description of the organic remains of the strata of the Tilgate Forest', in which his observations and sources of evidence are described in detail. This period is also dealt with in the two biographies of Mantell. See Spokes, *Gideon Algernon Mantell*, pp. 25–30. It is Professor Dennis Dean who describes Mantell's 1827 book as the 'rarest and most historic dinosaur book in English'; for his discussion of Mantell's contribution to the understanding of fossil reptiles at this point, see Dean, *Gideon Mantell*, pp. 89–96. Mantell's journal entries give a vivid insight into his practice: 'The Journal of Gideon Mantell', unedited, unpublished version in four volumes (Sussex Archaeological Society, Lewes, Sussex), Vol. 1, pp. 126, 130, 131, 135; Vol. 2, p. 15.

William Buckland's description of Mary Anning's first pterodactyl and details of the dramatic discovery appear in his paper, 'On the Discovery of a new species of Pterodactyle in the Lias at Lyme Regis', *Transactions of the Geological Society of London*, 2nd series, vol. 3 (1829), pp. 217–22. See also Lang, 'Mary Anning and the Pioneer Geologists'. Cuvier's interpretations are in G. Cuvier, *Recherches sur les Ossemens Fossiles* (Paris, 1824), Vol. 5, part 2, pp. 378–80.

The Annings' difficulties in selling fossils in the late 1820s and the changing fortunes of the Duke of Buckingham are described in M. A. Taylor and H. S. Torrens, 'Saleswoman to a new Science' in *Proceedings of the Dorset Natural History and Archaeological Society*, vol. 108 (1987). Mary Anning's arduous work is outlined in a letter to Mrs Murchison and can be found in W. D. Lang, 'Three Letters by Mary Anning',

Proceedings of the Dorset Natural History and Archaeological Society, vol. 66 (1944), p. 170. For the success of Henry de la Beche's print 'Duria Antiquior' see the journal *All the Year Round*, a report entitled 'Mary Anning, the Fossil Finder', vol. 13 (1865), pp. 60–3.

CHAPTER 8

Charles Lyell's masterpiece, *Principles of Geology*, was published in three volumes between 1830 and 1833. For a discussion on valley formation see p. 111 in the Penguin Classics edition, 1997, which includes an introduction by James Secord; for Lyell's views on parsons in England and the Mosaic account of Creation see p. xxiv. A full discussion of Lyell's theories can be found in Rudwick, *Meaning of Fossils*. The controversy between the fluvialists is also outlined in Rupke, *Chain of History*; for Conybeare's letter to Buckland describing Scrope as a goose see p. 86. For details of Lyell's correspondence with Mantell as the controversy over the Deluge wore on, see Lyell's sister-in-law's book: Lyell, *Life, Letters and Journals*, Vol. 1, pp. 252–3; letters dated April and June 1829.

Charles Lyell's letter to Mantell urging him to take the lead in fossil reptiles can be found in the New Zealand archives: Charles Lyell to Gideon Mantell, 23 Mar. 1829, Mantell MSS, ATL-NZ Folder 62, letter 55. I am very grateful to Alan John Wennerbom for permitting me to cite from his fascinating thesis on the relationship between Lyell and Mantell. Mantell's extensive preparations for his museum are evident from his diary; see Curwen, *Journal of Gideon Mantell*. Robert Bakewell's visit and preview are outlined in 'A visit to the Mantellian Museum at Lewes', *Natural History Magazine*, vol. 3 (1829), pp. 9–17, and are also discussed in Spokes, *Gideon Algernon Mantell*, pp. 34–5.

Gideon Mantell's landmark paper describing the Age of Reptiles is 'The Geological Age of Reptiles', *Edinburgh New Philosophical Journal* vol. II (Apr.–Oct. 1831), pp. 181–5. Mantell's diary shows that this paper was written two years earlier, on 3 November 1829, well before the plate 'Duria Antiquior', showing ancient Dorset seas teeming with reptilian life, was printed. The clergyman who objected to the very idea of an Age of Reptiles is described in Spokes, *Gideon Algernon Mantell*, p. 44. See also the *Bridgewater Treatise* by the Revd William Kirby (London: William Pickering, 1835), Vol. 1, pp. 36–42.

The first reference to Gideon Mantell as a British Cuvier was in a letter from Robert Bakewell to his publisher, Thomas Longman, cited in Dean, *Gideon Mantell*, p. 116. The reference to Mantell's 'genius' is to be found in a letter from Charles Lyell to J. Fleming, 7 Jan. 1835; see Lyell, *Life, Letters and Journals*, Vol. 1, p. 446.

Details of the royal trips to Brighton and Lewes are recorded in the local papers of the time; see 'Royal visit to Lewes: the Reception of William IV' at the Sussex Archaeological Society, ref. 942.25./Lew. More details of the royal family are in C. Fox, *London – World City 1800–1840* (Yale, 1992), and *The Royal Pavilion: the Palace of George IV*, published by Brighton Arts and Leisure Services. Mantell's diary also describes details of the event.

CHAPTER 9

Richard Owen's difficulties with his future mother-in-law are described in his grandson's biography: Owen, *Richard Owen*, pp. 35–45; for background to the unfolding saga of the prolonged engagement and Richard's and Caroline's response to it, see pp. 35, 37, 42, 60, 62, 63, 67, 89, 90. Further details can be found in British Library Add. MSS 39955, f. 212; and British Library Add. MSS 39955, f. 218. Their troubled relationship is also described in Desmond, *Politics of Evolution*, p. 250.

The clash between Geoffroy Saint-Hilaire and Georges Cuvier leading up to Cuvier's death is discussed in Schneer, *Towards a History of Geology*, ch. 2, by Frank Bourdier. See also Rudwick, *Meaning of Fossils*, ch. 3. Geoffroy's papers claiming that fossil animals gave rise to living ones is in Geoffroy Saint-Hilaire, 'Recherches sur l'organisation des gavials . . . et sur cette question, si les gavials . . . descendent, par voie non interrompue de génération, des gavials antédiluvians . . .', *Mémoires du Muséum d'Histoire Naturelle*, vol. 2 (1825), pp. 97–156. See also the same journal, vol. 17 (1828), pp. 209–30

I am indebted to Adrian Desmond for his colourful perspective on Robert Grant's career and the clash with Owen, outlined in Desmond, *Politics of Evolution*, pp. 8–11, 239 onwards. This is also discussed in Desmond's *Archetypes and Ancestors* (London: Blond & Briggs, 1982); see pp. 115–22. For Wakley's quotation in praise of Grant, see the *Lancet,* vol. 1 (1834), pp. 688–9.

Richard Owen's paper that launched him on to the scientific stage is 'Memoir on the Pearly Nautilus', published by 'direction of the Council' of the Royal College of Surgeons in 1832. The interest in monotremes to determine the existence of intermediate forms is discussed in Rupke, *Richard Owen*, pp. 77–9. A more detailed account is given in Desmond, *Politics of Evolution*, pp. 279–88. See also Owen's grandson's biography: Owen, *Richard Owen*, pp. 60–5.

Buckland's unusual domestic circumstances are described in Gordon, *Life and Correspondence*, pp. 90–110. See also his son Francis Buckland's 'Memoir of the Author', which appears in the 1858 edition of Buckland's *Bridgewater Treatise*; see p. xxxvi for an account of his parents writing together. Quotations given in this chapter from the *Bridgewater Treatise* are from pp. 8–11, 19–22, 125 onwards.

Nicolaas A. Rupke in *The Great Chain of History* has provided a fascinating perspective on the impact of the *Bridgewater Treatises*. Specialist sources cited in this section include, Hack, 'Geological Sketches' (1832), pp. 38–9, quoted in Rupke, on p. 167; and George Croly, *Blackwood's (Edinburgh) Magazine*, vol. xlii (1837), p. 690 and quoted in Rupke, p. 216. For the concerns over the relations between carnivorous animals and sin see George Bugg, *Scriptural Geology* (1826) Vol. I, pp. 139–45, 118. Thomas Thompson wrote for the *Magazine for Natural History* an article called 'An Attempt to Ascertain the Animals Designated in the Scriptures by the Names of Leviathan and Behemoth', vol. viii (1835), p. 320. The review of this theory can be found in the *Edinburgh New Philosophical Journal,* vol. 19 (1835), pp. 263–81.

For further discussion of this topic, see Schneer, *Towards a History of Geology*, and Milton Millhauser, 'The Scriptural Geologists', *Osiris*, vol. 11 (1954), pp. 65–86.

Richard Owen's success in claiming the zoological specimens for dissection is clearly shown in Owen, *Richard Owen*; see pp. 92, 95–6, 101, 106–7, 122, 169. Robert Grant's difficulties, leading to his eventual decline, are discussed in Desmond, *Politics of Evolution*, p. 291 onwards, and in his *Archetypes and Ancestors* (London: Blond & Briggs, 1982), pp. 42–4, 115 onwards. See also the *Lancet*, vol. 1 (1836–7), pp. 766, 21; also J. Beddoe, *Memories of Eighty Years* (Bristol: Arrowsmith, 1910), pp. 33.

The role of the newly formed British Association for the Advancement of Science as a vehicle for Owen's ambition, and the influence of his father-in-law in the ensuing battles with Mantell over fossil reptiles, are discussed in H. Torrens, 'Politics and Palaeontology: Richard Owen and the Invention of Dinosaurs', in J. Farlow and M. K. Brett-Surman, *The Complete Dinosaur* (Indianapolis: Indiana University Press, 1997), p. 174 onwards. See also J. Morrell and A. Thackray, *Gentlemen of Science: The Early Years of the British Association for the Advancement of Science* (Oxford: Clarendon Press, 1981), pp. 95 onwards, 31.

CHAPTER 10

John Cooper, the Curator of the Booth Museum, Brighton, and his team have searched the local archives for records of Mantell's museum in Brighton, and I am indebted to Cooper's generosity for the rich supply of original sources documenting Mantell's life in the mid- to late 1830s. Many of the citations are from the *Brighton Herald* and the *Brighton Gazette* between the years 1833 and 1838. These provide a vivid snapshot of Mantell's changing fortunes and the increasingly desperate attempts to keep his museum in Brighton. Another excellent guide to Mantell's difficulties can be found in a study by R. J. Cleevely and S. D. Chapman, 'The accumulation and disposal of Gideon Mantell's fossil collections and their role in the history of British Palaeontology', *Archives of Natural History* (1992), pp. 306–60.

This period in Mantell's life is also discussed in Spokes, *Gideon Algernon Mantell*, an account that includes the references from his correspondence with Professor Silliman. Some of the personal quotes that illustrate the optimism when embarking on the project and the severe depression which accompanied its failure are from Mantell's diary, Curwen, *Journal of Gideon Mantell*. See also Mantell's unpublished diary held at the Sussex Archaeological Society, Lewes, Vol. 2. More recently, Mantell's attempts to establish himself in Brighton are described in Dean, *Gideon Mantell*.

Mantell's interpretation of the Maidstone *Iguanodon* is analysed in D. B. Norman, 'Gideon Mantell's "Mantel-piece": the earliest well preserved Ornithischian Dinosaur', *Modern Geology,* vol. 18 (1993), pp. 225–45. For Bensted's account see W. H. Bensted, *The Iguanodon: Selections from the contributions to the Amici* (1836), pp. 70–7. The value of the Maidstone *Iguanodon* is considered in many publications; see T. Gardom and A. Milner, *The Natural History Museum Book of Dinosaurs* (Carlton, 1993), p. 93. Mantell's size comparisons with the iguana and his view of *Iguanodon* at this point are to be found in his *Geology of South-East England* (London: Thomas Longman, 1833), p. 312 onwards.

Records of the departure of Mrs Mantell are few, especially as Gideon Mantell virtually stopped writing in his diary during the turbulent years of 1837–40. According to Curwen, *Journal of Gideon Mantell*, p. 141, Mrs Mantell did not accompany her husband to Clapham. More recently, Dean, *Gideon Mantell*, p. 175, cites evidence that Mary Mantell finally departed on 4 March 1839, a few months after Mantell took on his London practice. In America, Professor Silliman later deleted all references to her departure in Mantell's letters before donating them to the Yale collection. It is thought that Walter Mantell, in accordance with his father's wishes, wiped the record from Mantell's journal.

CHAPTER 11

Professor Hugh Torrens at Keele University first showed that Richard Owen did not coin the term 'dinosaur' in August 1841, as was widely believed, but later, as he rewrote his Plymouth report for publication in April 1842. While writing this book, I enjoyed many discussions with Hugh Torrens about the likely sequence of events and the key important insights behind Owen's famous classification. His analysis is outlined in two articles: H. Torrens, 'Politics and Palaeontology: Richard Owen and the Invention of Dinosaurs', in Farlow and Brett-Surman, *Complete Dinosaur*, pp. 173–91; and for a more condensed view, H. Torrens, 'When did the Dinosaur get its name?', *New Scientist* Vol. 134, no. 1815 (4 Apr. 1992), pp. 40–4.

I am also grateful to Dr Angela Milner and Sandra Chapman of the Department of Palaeontology at the Natural History Museum in London, for guidance on the significance of Owen's skilled anatomical insights into the dinosaur's fused sacrum and mammal-like limb bones.

For information on the grant committee of the BAAS, see *The Report of the BAAS* (the 1837 meeting in Liverpool) (London: John Murray, 1838), Vol. 7; for the report on grants to geology see p. xix. See also *The Report of the BAAS* (the 1838 meeting) (London: John Murray, 1839), Vol. 8; see p. xxviii for details of the committee and p. xxx for the rules of grants. The way that the BAAS was hijacked by the London elite is described also in M. A. Taylor, 'The Plesiosaur's Birthplace . . .', *Zoological Journal of the Linnean Society*, vol. 112, (1994), pp. 179–96. Useful background on the formation of the BAAS and leading lights within the organisation is revealed in early letters: see J. Morrell and A. Thackray, *Gentlemen of Science: The Early Correspondence of the BAAS* (London: Royal Historical Society, 1984). Owen's patronage by leaders of the BAAS is also discussed in Rupke, *Richard Owen*. See also the letters from Sir Philip Egerton to Richard Owen at the Natural History Museum: especially 26 Oct. 1840. Owen's interest in the 'Enaliosauria' and his rivalry with Geoffroy Saint-Hilaire are outlined in Desmond, *Politics of Evolution*; see p. 324 onwards.

The excursion to visit Mary Anning and Thomas Hawkins is described in Owen, *Richard Owen*, p. 166. Hawkins's colourful character and background are to be found in Lang, 'Mary Anning and the Pioneer Geologists'. Mary Anning's frustration at having been exploited by the gentlemen of science is documented by William Lang in 'Mary Anning and Anna Maria Pinney', *Proceedings of the Dorset Natural History and Archeological Society*, vol. 80 (1959); it should be noted that some historians have suggested that as Pinney was still young, she may not be reliable testimony. Mary Anning's difficulties are also discussed by Lang in 'Mary Anning of Lyme, Collector and Vendor of Fossils', *Natural History Magazine* vol. 5, no. 34 (1936), pp. 64–81. See also quotations in Charles Dickens's journal *All the Year Round*, vol. 13 (1865), pp. 60–3. The issue of her deepening financial problems is analysed in M. A. Taylor and H. S. Torrens, 'Saleswoman to a new Science' in *Proceedings of the Dorset Natural History and Archaeological Society* Vol. 108 (1987). Information on her misfortune in entrusting her life savings to a conman is from Hugh Torrens (personal correspondence).

The anecdotes concerning Owen preparing his two reports on British fossil reptiles for the BAAS, including his impressions of travelling by rail, the different museums, and entertaining Buckland and Mantell, can be found in Owen, *Richard Owen*. More information on Holmes can be found in John Cooper's 'George Bax Holmes and his relationship with Gideon Mantell and Richard Owen', *Modern Geology*, vol. 18 (1993), pp. 183–208.

Mantell's circumstances in the late 1830s and early 1840s are outlined in the two biographies: Spokes, *Gideon Algernon Mantell*, pp. 105–20, and Dean, *Gideon Mantell*. His Brighton lectures formed the basis of his book *The Wonders of Geology* (London: 1838). According to Dean, George Richardson, Mantell's Curator at Brighton, transcribed the Brighton lectures and these transcriptions were used to compile the first draft, which Mantell later revised. Although Richardson was credited as 'editor' in the first two editions, his name was later dropped.

For an analysis of the development of Murchison's work on the Silurian system and his clash with Sedgewick over the Cambrian, and with de la Beche over the Devonian, see Rudwick, *Meaning of Fossils*. See also Geikie, *Founders of Geology*.

The report of the eleventh meeting of the British Association for the Advancement of Science, held at Plymouth in August 1841 (and published in London by John Murray in 1842), provides a wealth of information about the evidence available to Owen, the collections he had seen, his interpretations, and the way Mantell's finds were used. Owen's lengthy 'Report on British Fossil Reptiles' is on pp. 60–201; he describes *Poekilopleuron* on pp. 84–8; *Streptospondylus* on pp. 88–94; *Cetiosaurus* on pp. 94–103; *Megalosaurus* on pp. 103–10; *Hylaeosaurus* on pp. 111–20; *Iguanodon* on pp. 120–144; and his fascinating summary is on pp. 191–201. Owen's summary chart, in which numerous reptiles appear with his own name beside them and *Iguanodon* is presented as though discovered solely by Cuvier, is on p. 190. Whewell's talk, and details of BAAS accounts and former presidents, etc., are also to be found in this volume.

The *Literary Gazette* for 14 Aug. 1841 provides a full summary of Owen's BAAS talk. For Mantell's reply in that same journal see 28 Aug. 1841, pp. 556–7. The significance of Owen's battle with the evolutionists in his report is considered by Stephen Jay Gould in 'An Awful, Terrible Dinosaurian Irony', *Natural History Magazine*, vol. 107, no. 1 (1998), p. 24 onwards. The increasing size estimates of *Iguanodon* as interpreted by Holmes are described in *Horsham, its History and Antiquities* (London: William Macintosh, 1868), p. 225 onwards.

Mantell's own report for the Royal Society in 1841 is 'A Memoir on . . . the Iguanodon and on the Remains of the Hylaeosaurus and other Saurians discovered in the Strata of the Tilgate Forest', *Philosophical Transactions of the Royal Society*, part 2, pp. 131–52. In this paper Mantell observes that *Iguanodon* 'with its long, slender prehensile fore-feet was enabled, while supported by its enormous hinder limbs, to pull down and feed on the foliage and trunks of the Clathrariae, Dracaenae Yuccae and arborescent ferns'. See also 'On the Fossil Remains of Turtles . . .' by G. A. Mantell in the same issue, pp. 153–8.

Mantell's carriage accident is described in his journal at the Sussex

Archaeological Society, Lewes; see Vol. 2. See also Curwen, *Journal of Gideon Mantell*. The effects of the subsequent illness, which affected him for the rest of his life, are outlined in correspondence with Silliman and cited in Spokes, *Gideon Algernon Mantell*, pp. 135–45, 251–60.

CHAPTER 12

Richard Owen's meteoric rise through Victorian society is reflected in Owen, *Richard Owen*, which records a succession of social events and meetings with leading figures of the day. Some of these are also discussed in Rupke, *Richard Owen*; see pp. 124–8 for the story of the large, flightless bird the moa, which illustrates how his ideas were received. The unnamed report showing that no work of Owen's created such excitement is cited in R. W. Clark, *Old Friends at Cambridge and Elsewhere* (London, 1900), p. 373. A further description of Owen's skilled identification of the moa can be found in Owen, *Richard Owen*, pp. 147–51.

Owen's own papers to the Zoological Society provide useful insights into this episode: 'Exhibition of a bone of an unknown struthious bird of large size from New Zealand', *Proceedings of the Zoological Society*, vol. 7 (1839), pp. 169–71; and 'Notice of fragment of the femur of a gigantic bird of New Zealand', *Proceedings of the Zoological Society*, vol. 3, (1842), pp. 29–32. Significant letters which shed light on Owen's reputation and success are in the British Library: see Broderip to Buckland, 20 Jan. 1843, BL Add. MSS 38091, f. 193. Dr Rule's contrasting version of events is given in Curwen, *Journal of Gideon Mantell*, p. 225.

Vestiges of the Natural History of Creation, which caused a sensation in 1844, was by a Scottish journalist called Robert Chambers. For a detailed discussion of the impact of this book see Gillispie, *Genesis and Geology*. Owen's theory of archetypes is outlined in *On the Archetype and Homologies of the Vertebrate Skeleton* (London, 1848), and *On the Nature of Limbs* (London, 1849). These ideas are discussed in Rudwick, *Meaning of Fossils*, and in Kevin Padian's 'The rehabilitation of Sir Richard Owen', *BioScience*, vol. 47, no. 7 (1997).

Mantell's attempts to rebuild his life in Clapham are described in many sources. His own diary and his correspondence with Silliman, cited at length in Spokes, *Gideon Algernon Mantell*, illustrate his personal circumstances vividly. Mantell's resentment of Owen from 1842 is revealed by his correspondence and by biographical material. The antagonism between the two men became public following Mantell's disagreement with Owen over the interpretation of the belemnites. This and other aspects of their feud are analysed by D. T. Donovan and M. D. Crane, 'The Type Material of the Jurassic Cephalopod Belemnotheutis', *Palaeontology*, vol. 35 (1992), pp. 273–96.

The difficulties faced by Mary Anning in her final years were first outlined by William Lang, 'Mary Anning of Lyme, Collector and Vendor of Fossils', *Natural History Magazine*, vol. 5, no. 34 (1936), pp. 64–81. See also W. Lang, *Proceedings of the Dorset Natural History and Archaeological Society*, including 'Mary Anning and the Pioneer Geologists'; and 'More about Mary Anning', vol. 76 (1956), for the story of Nellie Waring's visit to the shop. A more recent summary can be found in H. Torrens, 'Mary Anning of Lyme; the greatest fossilist the world ever knew', *British Journal of the History of Science*, vol. 28 (1995), pp. 257–84. Charles Dickens's journal *All the Year Round*, vol. 13 (1965), pp. 60–3, credits the 'carpenter's daughter'. I am grateful also to the Reverend Thomas Goodhue of the Long Island Council of Churches in New York for drawing to my attention the entries she made in her commonplace book before her death, as described in his talk at the Anning bicentennial conference in Lyme in 1999.

Mantell's efforts to establish a second collection are revealed by his diary entries. The significance of this second collection is analysed in S. D. Chapman and R. J. Cleevely, 'The Accumulation and Disposal of Mantell's Fossil Collections', *Archives of Natural History* (1992). For Walter's fossil contribution see J. C. Yaldwin, G. Tee and A. Mason, 'The status of Gideon Mantell's "first" *Iguanodon* tooth in the Museum of New Zealand, Te Papa Tongarewa', *Archives of Natural History*, vol. 24, no. 3 (1997), pp. 397–421. The fate of Mantell's Brighton Curator, George Richardson, is portrayed vividly by Hugh Torrens and John

Cooper, 'George Fleming Richardson: Man of letters, lecturer and Geological Curator', *Geological Curator*, vol. 4, no. 5 (1985), issue 2, pp. 249–68.

Mantell's later work attempting to elucidate further details of *Iguanodon* is described in two papers in *Philosophical Transactions of the Royal Society*: 'On the Star Jaws of the Iguanodon' (1848), and 'Additional Observations on the Osteology of the Iguanodon and Hylaeosaurus' (1849). Mantell's work has been analysed by Dr David Norman at Cambridge University in 'Gideon Mantell's "Mantel-Piece": the earliest well preserved Ornithischian Dinosaur', *Modern Geology*, vol. 18 (1993), pp. 225–45. The conflicts with Richard Owen arising through the collector George Holmes are analysed in John Cooper's 'George Bax Holmes and his relationship with Gideon Mantell and Richard Owen', *Modern Geology*, vol. 18 (1992), pp. 183–208. For Mantell's discovery of additional dinosaurs, see Spokes, *Gideon Algernon Mantell*, and Curwen, *Journal of Gideon Mantell*. The finds are also described by Mantell in *Petrifactions and their Teachings* (London: Bohn, 1851), pp. 224–5, 330–2; see also Mantell's *Geological Excursions around the Isle of Wight* (London: H. G. Bohn, 1854), pp. 332. His discoveries are discussed by Dean, *Gideon Mantell*, pp. 236–9. The history of the discovery of sauropods is given in John. S. McIntosh, M. K. Brett-Surman and James O. Farlow, 'The Discovery of Sauropods', in Farlow and Brett-Surman, *Complete Dinosaur*, pp. 264–89.

A compelling profile of Owen's character can be found in the opening chapters of Adrian Desmond's *Archetypes and Ancestors* (London: Blond & Briggs, 1982). Owen as a 'social experimenter with a penchant for sadism and mystification' is in W. Irvine, *Apes, Angels and Victorians* (New York: McGraw-Hill, 1955); see also p. 181 for his clash with Hugh Falconer. This is also outlined in Mantell's diary and in Kevin Padian's 'The rehabilitation of Sir Richard Owen', *BioScience*, vol. 47, no. 7 (1997). Owen as a man 'addicted to acrimonious controversy' is described in W. H. Flowers's profile of Richard Owen in the *Dictionary of National Biography*, pp. 1329–39. He is described as driven by arrogance and jealousy in Gavin de Beer, *Charles Darwin, Evolution by Natural*

Selection (New York: Doubleday, 1964). For details of Huxley's struggle to establish himself in London and his despair of finding a scientific post, see L. Huxley, *Life and Letters of Thomas Henry Huxley* (London: Macmillan, 1903), chs 5, 6.

CHAPTER 13

Information on the Great Exhibition of 1851 can be found in D. Newsome, *The Victorian World Picture* (London: John Murray, 1997). This account includes a description of the procession and of the reaction of the Archbishop of Canterbury. *The Letters of Queen Victoria*, Vol. II, pp. 316–19, show her delight in the Exhibition, as do her journal entries for May 1851. See also C. H. Gibbs-Smith, *The Great Exhibition of 1851* (London: Victoria & Albert Museum 1950); this account lists the jurors and describes some of the exhibits. For more information see R. E. Prothero, *The Life and Correspondence of Arthur Penrhyn Stanley*, Vol. 1 (1893), pp. 200–5. Caroline Owen's account of the procession is in Owen, *Richard Owen*, Vol. 1, p. 366. Mantell's descriptions can be found in his journal for May and the ensuing months.

Buckland's tragic descent into madness is discussed in many sources. The most vivid is J. W. Gruber and J. C. Thackray, *Richard Owen Commemoration* (London: Natural History Museum Publications, 1992), pp. 78–80. The concern that Buckland was lying 'at the Clapham retreat, amongst outrageous madmen' was expressed by Thomas Hawkins in a letter to Richard Owen, 16 May 1851; see the Owen Collection at the Natural History Museum 14:516/7. The view that 'the mental disease will take such a root that nothing will remove it' is in a letter from Murchison to Owen, 25 Jan. 1850, and cited in Gruber and Thackray (above).

For a detailed analysis of the *Telerpeton/Leptopleuron* dispute between Owen and Mantell, see M. J. Benton, 'Progressionism in the 1850s: Lyell, Owen, Mantell and the Elgin fossil reptile *Leptopleuron* (*Telerpeton*)', *Archives of Natural History*, vol. 11, no. 1 (1982), pp. 123–36. Benton makes the point that Owen may not, in fact, have acted badly on this occasion. Patrick Duff had sent Owen the report of the discovery in the *Elgin Courant* and some drawings well before Mantell saw the specimen, and Owen believed that he was requested to describe the animal. The report which described Owen as 'prigging Mantell's bones' and suggested that Owen 'worrited Mantell to death' is anonymous: *A Sad Case . . .*, outlined in *Public Opinion* (1863), p. 490 onwards.

Details of the Crystal Palace Board's request to Mantell to construct a geological court are in the extracts from minutes of a meeting of the Board of Directors, 10 Aug. 1852, and can be found in the Alexander Turnbull Library, Wellington, New Zealand, MS papers 0083–032. I am grateful to Professor Hugh Torrens for alerting me to these details; the issue is discussed in full in his chapter, 'Politics and Palaeontology: Richard Owen and the Invention of Dinosaurs', in Farlow and Brett-Surman, *Complete Dinosaur*, p. 187. The deterioration in Mantell's health at this point is evident from his diary and the account in Spokes, *Gideon Algernon Mantell*.

The anonymous obituary of Gideon Mantell attributed to Richard Owen is in the *Literary Gazette* for 13 Nov. 1852, p. 842. For correspondence prompted by this see W. Hopkins to L. Horner, 17 Nov. 1852, IC 18.228, and W. Hopkins to E. Forbes, 4 Dec. 1852, IC 18.224. This is also discussed in Adrian Desmond, *Archetypes and Ancestors* (London: Blond & Briggs, 1982), p. 208. There are numerous other notices and obituaries: see especially the *Athenaeum*, 20 Nov. 1852, pp. 1270–1; *Literary Gazette*, 27 Nov. 1852, p. 1; *The Gentleman's Magazine*, vol. 38 (Dec. 1852), pp. 644–7; the anniversary meeting of the Royal Society on 30 Nov. 1852, 'Address by the Right Honourable Earl Rosse', *Quarterly Journal of the Geological Society* (1853); the anniversary address by William Hopkins, in the same journal, pp. xxii–xxvi; T. G. Vallance, 'Gideon Mantell: a Focus for Study in the History of Geology . . .', *Hist.Sci.New Zealand* (Feb. 1983). Many of the obituaries are summarised in Dean, *Gideon Mantell*, pp. 264–6.

The magnificent banquet in the belly of the *Iguanodon* is described in many sources. A good summary of the event is in B. C. Gardiner, 'Clift,

Darwin, Owen and the Dinosauria', *The Linnean* (1991), pp. 19–27. For interpretations of dinosaurs at the time and the public reaction, see 'The Fossil Dinner', *London Quarterly Review*, vol. 3, no. 5 (1854), pp. 232–79. The dinner is also described in 'The Crystal Palace at Sydenham', *Illustrated London News*, vol. 23, no. 661 (31 Dec. 1853), pp. 599–600; and in the same journal, vol. 24, no. 662 (7 Jan. 1854), p. 22. See also S. McCarthy and M. Gilbert, *The Crystal Palace Dinosaurs* (London: The Crystal Palace Foundation, 1994).

Preparations for the Crystal Palace dinosaurs are described in *The Times* of 2 Nov. 1853, and see also the *Crystal Palace Herald*, vol. 1, no. 1 (1853). The reaction of the public is shown in the *Westminster Review* vol. 62, no. 6 (1854), pp. 540–1, and in L. Figuier, *The World before the Deluge* (London: Cassell, Petter & Galpin, 1863), pp. viii, 449 onwards. Fictional accounts include Jules Verne, *Journey to the Centre of the Earth* (Paris: Hetzel, 1864), pp. 2, 335 onwards; Charles Dickens, *Bleak House* (London, 1853). Richard Owen's account is in *Geology and the Inhabitants of the Ancient World* (London: Crystal Palace Library, 1854); see also Benjamin Waterhouse Hawkins, *Crystal Palace. Guide to the Palace and Park* (London: Dickens & Evans, 1877), and 'On visual education as applied to geology', see T. Hawkins in *Journal of the Society of Arts*, vol. 2. pp. 444–9.

The fate of Mantell's sons and of his collections is discussed in J. C. Yaldwin, 'The status of Gideon Mantell's "first" Iguanodon tooth . . .', *Archives of Natural History*, vol. 24, no. 3 (1997), pp. 397–421. See also Spokes, *Gideon Algernon Mantell*.

For Professor Owen's ambitious plans for a national museum, see Owen, *Richard Owen*, Vol. 2. His museum plans are also discussed in detail by Rupke, *Richard Owen*.

CHAPTER 14

Huxley's campaign to curb Owen's power and block his ambition for a natural history museum is discussed in Rupke, *Richard Owen*, pp. 98–102. See also 'Report from the select committee', *Parliamentary Papers 1860* (540), vol. 16, p. 303, for Huxley's comments on Owen's 'little matured' scheme.

There are numerous sources discussing the development of Darwin's ideas. The most compelling biography is Desmond's and Moore's *Darwin*. I have leaned heavily on this account for Darwin's personal reactions, especially his fears over how his ideas might be received and the reaction of Professor Owen. For a general account of the development of Darwin's thinking see also William Irvine's *Apes, Angels, and Victorians: The Story of Darwin, Huxley and Evolution* (New York: McGraw-Hill, 1955).

The quotation on natural selection given at length in this chapter is from Darwin's *Origin of Species by means of Natural Selection* (London: John Murray, 1859), p. 61. For the influence of Lyell on Darwin, see F. Burkhardt and S. Smith (eds), *Correspondence of Charles Darwin*, Vol. 3 (1987), p. 55: Darwin to L. Horner (1844). For Thomas Huxley's response see Leonard Huxley, *Life and Letters of Thomas Henry Huxley* (London and New York: Macmillan, 1903), pp. 258, 239, 254. Owen's review, which caused Darwin so much anxiety, is in the *Edinburgh Review or Critical Journal* (Jan.–Apr. 1860), pp. 487–532; quotations given here are from pp. 494, 502, 521, 500. The report in the *Manchester Spectator* can be found in Owen's file at the Natural History Museum, BM (NH), L, OC 18, dated 8 and 22 Dec. 1849. Rupke, *Richard Owen*, discusses Owen's response to Darwin on pp. 232–42. The differences in ideology between Darwin and Owen are analysed in J. W. Gruber and J. C. Thackray, *Richard Owen Commemoration* (London: Natural History Museum Publications, 1992), see pp. 71–81.

Numerous stories have built up around the famous clash between Wilberforce and Huxley in 1860 at the BAAS. I have drawn from the accounts in Leonard Huxley, *Life and Letters of Thomas Henry Huxley* (above), see pp. 259–74; Leonard Huxley quotes many different versions written by friends to Darwin. This meeting is also discussed in Desmond and Moore, *Darwin*, pp. 492–9 and in William Irvine's, *Apes, Angels and Victorians: The Story of Darwin, Huxley and Evolution* (above), pp. 1–8.

For a full discussion of the ape brain controversy see Rupke, *Richard Owen*, ch. 6 on 'Cerebral Constructs'. Huxley's own view of events is described in Leonard Huxley's *Life and Letters of Thomas Henry Huxley* (above), see p. 277. Thomas Huxley presented his arguments in *Evidence as to Man's Place in Nature* (London: 1863). The way Huxley outmanoeuvred Owen at the Zoological Society and the Royal Society is summarised in Desmond and Moore, *Darwin*, p. 505. One of several caricatures of Owen at this time appeared in *A Sad Case, Recently Tried before the Lord Mayor, Owen vs Huxley'*, otherwise known as the 'Bone Case'; it was published in London in 1863, and the author is thought to have been George Pycroft. The *Punch* poem appears in the issue of 15 May 1861, under a picture of a gorilla bearing the sign, 'Am I a Man and a Brother?'

Huxley's studies placing the *Dinosauria* in an evolutionary context are outlined in several sources. A vivid account which shows the development in Huxley's thinking can be found in Adrian Desmond's *Huxley* (Penguin, 1998), pp. 299–300, 356–60. See also Desmond, *Archetypes and Ancestors* (London. Blond & Briggs, 1982), p. 124 onwards, for the debate on the significance of the *Archaeopteryx*. For the quotations from letters to Lyell at the Royal Institution, see Leonard Huxley's *Life and Letters* (above), pp. 381, 424. A brief summary of Huxley's classification is presented in E. H. Colbert, *Men and Dinosaurs* (Penguin, 1968). See also Farlow and Brett-Surman, *Complete Dinosaur*, ch. 39.

Owen's reluctance to accept evolution as outlined by Darwin and his enthusiasm for retaining Divine laws are described in J. W. Gruber and J. C. Thackray, *Richard Owen Commemoration* (London: Natural History Museum Publications, 1992). The sad fate of his son William is also described there (see pp. 81–2). The dispersal of some of Mantell's collection while Owen was in charge of the British Museum is discussed in S. D. Chapman and R. J. Cleevely, 'The accumulation and dispersal of Gideon Mantell's fossil collection and their role in the history of British palaeontology', *Archives of Natural History*, vol. 19, no. 3 (1992), pp. 307–64.

Owen's being 'systematically written out of history' by Darwin's supporters is described in Rupke, *Richard Owen*, pp. 3–5. The young Regius Professor of Modern History at Oxford who described Owen as 'a bad case' is quoted in Gruber and Thackray, *Richard Owen Commemoration* (above), p. 4.

Select Bibliography

Buckland, William, 'Notice on the Megalosaurus or Great Fossil Lizard of Stonesfield', *Transactions of the Geological Society of London*, vol. 2, no. 1 (1824), pp. 390–7.

Buckland, William, *Geology and Mineralogy Considered with Reference to Natural Theology*, 2 vols (London: W. Pickering, 1837).

Curwen (ed.), E. C., *The Journal of Gideon Mantell, Surgeon and Geologist: 1818–1852* (London: Oxford University Press, 1940).

Dean, Dennis, *Gideon Mantell and the Discovery of Dinosaurs* (Cambridge University Press, 1999).

Desmond, Adrian, *The Politics of Evolution: Morphology, Medicine, and Reform in Radical London* (Chicago and London: University of Chicago Press, 1989).

Desmond, Adrian, and Moore, James, *Darwin* (London: Penguin, 1992).

Farlow, James O., and Brett-Surman, M. K., *The Complete Dinosaur* (Indianapolis: Indiana University Press, 1997).

Geikie, Archibald, *The Founders of Geology* (London: Macmillan & Co., 1897), reprinted in 1962, New York: Dover Publications.

Gillispie, Charles Coulston, *Genesis and Geology: A Study in the Relations of Scientific Thought, Natural Theology and Social Opinion in Great Britain, 1790–1850* (Cambridge, Mass.: Harvard University Press, 1951).

Gordon, Anna B., *The Life and Correspondence of William Buckland* (London: John Murray, 1894).

Irvine, W., *Apes, Angels and Victorians: The Story of Darwin, Huxley and Evolution* (New York: McGraw-Hill, 1955).

Lang, William D., 'Mary Anning and the Pioneer Geologists of Lyme',

Proceedings of the Dorset Natural History and Archaeological Society, vol. 60 (1939), pp. 142–64.

Lyell (ed.), Katherine M., *Life, Letters and Journals of Sir Charles Lyell, Bart.*, 2 vols (London: John Murray, 1881).

Mantell, G. A., *The Fossils of the South Downs* (London: Lupton Relfe, 1822).

Mantell, G. A., 'Notice on the Iguanodon, a Newly discovered Fossil Reptile from the sandstone of the Tilgate Forest in Sussex', *Philosophical Transactions of the Royal Society of London*, vol. 115 (1825), pp. 179–86.

Mantell, G. A., *Illustrations of the Geology of Sussex* (London: Lupton Relfe, 1827).

Mantell, G. A., *Thoughts on a Pebble* (London: Reeve, Benham & Reeve, 1849).

Mantell, G. A., Journal in four volumes from 1819 to 1852 (unpublished); typed manuscript held at the Sussex Archaeological Society, Lewes, Sussex.

Owen, Richard, 'Report on British Fossil Reptiles', Part 2, *Report of the British Association for the Advancement of Science for 1841* (London: John Murray, 1842), pp. 60–204.

Owen (ed.), R. S., *The Life of Richard Owen*, 2 vols (London: John Murray, 1894).

Rudwick, Martin J. S., *The Meaning of Fossils* (Chicago: University of Chicago Press, 1972).

Rupke, Nicolaas A., *The Great Chain of History* (Oxford: Clarendon Press, 1983).

Rupke, Nicolaas A., *Richard Owen, Victorian Naturalist* (New Haven and London: Yale University Press, 1994).

Schneer, Cecil J., *Towards a History of Geology* (Boston, Mass.: MIT Press, 1967).

Spokes, Sidney, *Gideon Algernon Mantell, LLD, FRCS, FRS, Surgeon and Geologist* (London: John Bale & Sons & Danielson, 1927).

Torrens, H. S., 'The Dinosaurs and Dinomania over 150 years', *Modern Geology* vol. 18 (1993), pp. 257–86.

Index